W0081483

Researching Internet Governance

Information Policy Series

Edited by Sandra Braman

The Information Policy Series publishes research on and analysis of significant problems in the field of information policy, including decisions and practices that enable or constrain information, communication, and culture irrespective of the legal silos in which they have traditionally been located as well as state-law-society interactions. Defining information policy as all laws, regulations, and decision-making principles that affect any form of information creation, processing, flows, and use, the series includes attention to the formal decisions, decision-making processes, and entities of government; the formal and informal decisions, decision-making processes, and entities of private and public sector agents capable of constitutive effects on the nature of society; and the cultural habits and predispositions of governmentality that support and sustain government and governance. The parametric functions of information policy at the boundaries of social, informational, and technological systems are of global importance because they provide the context for all communications, interactions, and social processes.

Researching Internet Governance

Methods, Frameworks, Futures

Edited by Laura DeNardis, Derrick L. Cogburn,
Nanette S. Levinson, and Francesca Musiani

The MIT Press
Cambridge, Massachusetts
London, England

The Open Access edition of this book was published with the support of a generous grant from the Hewlett Foundation Cyber Initiative to the Internet Governance Lab at American University.

This book was set in Stone Serif and Stone Sans by Westchester Publishing Services. Printed and bound in the United States of America.

Library of Congress Cataloging-in-Publication Data

Names: DeNardis, Laura, 1966– editor. | Cogburn, Derrick L., editor. |
   Levinson, Nanette S., editor. | Musiani, Francesca, editor.
Title: Researching internet governance : methods, frameworks, futures /
   edited by Laura DeNardis, Derrick L. Cogburn, Nanette S. Levinson, and
   Francesca Musiani.
Description: Cambridge, Massachusetts : The MIT Press, 2020. |
   Series: Information policy series | Includes bibliographical references and index.
Identifiers: LCCN 2020000432 | ISBN 9780262539753 (paperback)
Subjects: LCSH: Internet governance—Research.
Classification: LCC TK5105.8854 .R47 2020 | DDC 384.3/34072—dc23
LC record available at https://lccn.loc.gov/2020000432

10 9 8 7 6 5 4 3 2 1

Contents

Series Editor's Introduction

Sandra Braman

Becoming aware of a new subject that needs research is an intellectual challenge in its own right. Making it legible to others and developing a research agenda are even more so. Building out a field, for those subjects of research so broad, complex, and important that they deserve or require it, is a yet greater challenge by orders of magnitude.

Brian Kahin gets credit for the first round of making the subject of Internet governance research legible and stimulating the development of research agendas. As Founding Director of the Harvard Information Infrastructure Project (1989–1997), he organized a series of influential conferences that produced, in turn, a series of edited or coedited books, including several with MIT Press: *Public Access to the Internet* (1995), *Standards Policy for Information Infrastructure* (1995), *Borders in Cyberspace* (1997), *Coordinating the Internet* (1997), and more. It was in Kahin's conferences that Michael Goldhaber first introduced the concept of the attention economy, Jeffrey MacKie-Mason (now, as University Librarian at the University of California-Berkeley, famously providing leadership in the area of open access and the economics of scholarly publishing in the digital environment) discussed "unbundling" journal articles for separate sale of the elements, access to the Internet received the scholarly attention it deserved, the importance of standards and protocols came to be appreciated by a much wider group of thinkers than the very small group of cognoscenti historically involved, and many of us working on what we might otherwise have perceived as disparate research topics came to see the relationships among them all.

Laura DeNardis gets credit for the second round, currently underway, that has completed the task of establishing the field. Her own books are foundational, always required reading, including those published by MIT Press: *Protocol Politics* and the edited *Opening Standards*. Going further, though, in leadership positions that began with service as Director of the Yale Law School Information Society Project and went on to include being Director of Research for the Global Commission on Internet Governance as well as formal and informal advisor to a number of nonprofit and governmental organizations—positions that have offered opportunities to fund and otherwise support and inspire research—DeNardis has systematically built out the field. This is the second book produced by the editorial team she has recently pulled together to expand on the effort of conceptualizing the domain, a team that includes Francesca Musiani of the CNRS in France and DeNardis's American University colleagues Nanette Levinson and Derrick Cogburn. The group's first book addressed the Internet as infrastructure.

My own thoughts regarding where the field of Internet governance is going can be found in my chapter in this book. Here, the point is other: what it takes to make an all-important subject of research visible, help scholars develop their research agendas, and build a field. This collection stands on its own, with great value for students and scholars. It also marks the "coming of age" of the field. The work does not claim to be comprehensive but, rather, to provide a sense of the range of the field, diverse ways of thinking about it, and examples of quite disparate types of research methods that can be used to study it.

Coeditor Levinson's concluding chapter contextualizes the whole within the history of the sociology of knowledge, providing insight into the processes of learning about Internet governance in a manner that should be inspirational for those considering joining this scholarly community as well as those who rely upon it in their own scholarship, research, and policy-making. The chapters by DeNardis and Mueller and Badiei, as well as my own, contextualize the field relative to its own history and other bodies of knowledge.

We see how the field looks from the perspective of the law (Weber), science and technology studies (coeditor Musiani), and information security (Deibert). There are exemplars of methods that range from text mining (coeditor Cogburn) and various types of technical analysis (Jardine; Deibert; and Hall, Madaan, and O'Hara) to discourse analysis (ten Oever, Milan, and Beraldo; as well as Hofmann) and interviews (Jørgensen).

The collaborative and multinational nature of the editorial team, as well as the book's authors and content, are indicators of the field's evolution. We are far from done thinking about how to think about Internet governance research, but this is a very strong place to begin.

1 Introduction: Internet Governance as an Object of Research Inquiry

Laura DeNardis

Why Study Internet Governance?

Governance of the Internet has quickly become one of the most pressing geopolitical issues of the contemporary era. How the Internet is designed and administered implicates a host of public policy concerns such as personal privacy, economic stability, national security, freedom of expression, and digital equality. Governments increasingly discuss issues of Internet governance and cybersecurity in the same breath as other types of global collective action problems such as terrorism, environmental protection, and human rights issues from poverty to child trafficking.

What was once an esoteric set of issues relegated to the technical community and a handful of scholars is now high on the policy agenda of all governments. The reasons for this escalation of interest in how the Internet is administered are absolutely clear. The economic stakes of the digital economy are immense, with all industry sectors dependent on the Internet to function and digital trade measured in trillions of dollars annually. An outage in cyberspace is an outage of the global economy. Internet policies also profoundly affect individual civil liberties and political discourses around elections. Governments have recognized that Internet governance has become a proxy for state power in areas ranging from cyber conflict to systems of filtering and censorship.

A number of Internet conflicts covered extensively in the media—such as Edward Snowden's disclosures about expansive government surveillance, massive data breaches, and Russian hacking during the 2016 US presidential election—have drawn public attention and scrutiny to questions about how the Internet is controlled and administered, whether by content intermediaries like social media companies, by traditional governments, or by new institutions designed to manage the security and stability of critical Internet resources.

Because of the high stakes of Internet governance questions, there is a tremendous interest in empirical scholarship that provides an evidence base for the decisions of policy makers and the private sector. There is equally a need for scholarship that makes visible to society the sinews of power constructing and controlling the Internet and explaining what the implications are for society and the economy.

There is also an epistemic community of scholars—highly interdisciplinary and distributed around the globe—that self-reflexively identifies as being a global Internet governance research community. The scholarly community has been increasingly organized, at least since the inception of the Global Internet Governance Academic Network (GigaNet) in 2006 and since the rise of interdisciplinary centers on Internet policy, cyber governance, Internet governance, and related initiatives at major universities around the world. Increasing numbers of graduate students, advocacy organizations, academic centers, policy makers, and new kinds of firms that recognize their own Internet policy challenges seek to better understand the choices and implications of how global digital networks are governed.

The defining and original feature of this book is that the topic is research concepts, methods, and frameworks. Numerous books (many by the authors in this book) contain state-of-the-art research *on* Internet governance topics, rather than viewing Internet governance research *as* the topic. How to begin to study Internet governance? This chapter examines the following questions: What is the thing studied when one studies Internet governance? What is the evidence base being examined? Who is studying Internet governance, and what methodologies and conceptual lenses are instructive? This chapter also lays out the organization of the book's chapters—contributed by leading scholars in fields as diverse as law, computer science, communication, science and technology studies, and political science. The rising stakes and increasing visibility of control struggles over Internet governance, as well as the coalescing and increasingly maturing of an interdisciplinary field that epistemically describes what it is doing as Internet governance research, present an important moment of opportunity for this volume.

What Is the Thing Studied When One Studies Internet Governance?

The Internet is and always has been governed, although not in the traditional sense of nation-state governance but in points of coordination and control that cross borders and are distributed among many actors, including the private sector, traditional governmental structures, new global institutions, and sometimes, citizens themselves. Governance is not only about governments. It is enacted via technical design, resource coordination, private ordering, and conflicts at control points.

Internet governance can be generally defined as the administration and design of the technologies that keep the Internet operational and the enactment of policy around these technologies. Beneath the things that humans perceive—content, applications, and devices—when using a network, there are thousands of behind-the-scenes control points. There is no one ideal taxonomy for describing these many points of design, coordination, and control, but one way to organize the functions is as follows (see DeNardis 2014; DeNardis and Musiani 2016):

- Administration of critical Internet resources such as names and numbers
- Establishment of Internet technical standards (e.g., protocols for addressing, routing, encryption, compression, error detection, identity systems, authentication)
- Coordination of access and interconnection (e.g., IXPs, net neutrality, access policies)
- Cybersecurity governance
- The policy-making role of private information intermediaries (e.g., via platform governance, algorithmic ordering, terms of service, computational ordering and decisions by artificial intelligence, and policies about security, speech, reputation, and privacy)
- Technical architecture-based intellectual property rights enforcement

These are all in themselves complex and multivariable points of control, and they overlap in many ways. Any of these tasks can be used to keep the Internet free and open; any can be exploited by governments or the private sector to enact censorship or carry out surveillance. This taxonomy, while capacious, actually bounds the scope of Internet governance as a target of research in important ways. For one, it clearly demarcates Internet governance from the enormous body of Internet studies focusing on how people, businesses, and governments use the Internet, what they say on the Internet (from the mundane to the political), or how user-centric questions such as identity and representation unfold. Those questions are the purview of the larger context of Internet research—obviously, not meaning how the Internet is used for research but meaning research about Internet usage (e.g., Markham and Baym 2008).

To oversimplify the distinction, whereas much of broader Internet research studies what people express on social media—such as content analysis of political expression on Twitter or issues of identity, representation for marginalized communities, or community formation—Internet governance research studies the mechanisms of control beneath the surface layer of content, such as algorithmic ordering, security, platform affordances, privacy policies instantiated in terms of service, mechanisms to detect fake bot accounts, and the regulatory contexts constraining or enabling all this. There is nothing natural about these distinctions, but there is pragmatic utility in bounding the scope of Internet governance.

As a prelude to interrogating Internet governance research, the following presents five distinguishing features about how the Internet is governed in practice: (1) Technical design and coordination decisions establish public policy. (2) Technologies of Internet governance, as currently designed, cross borders in a way that complicates nation-state jurisdiction. (3) Governance is distributed across multiple actors in a model usually described as private-sector-led, multistakeholder governance. (4) Internet security is both converging and diverging with national security. (5) Internet infrastructure control is now a proxy for political and economic power. Some of these serve as points of reference for this book's chapters, either as part of the conceptual framework or as concepts that are challenged and interrogated as Internet governance research enters its next stage.

Technical Design Enacts Governance

One distinguishing feature of the practice of Internet governance is that the design of technical architecture is a significant force enacting public policy. Therefore, a significant body of Internet governance research studies the underlying technologies of the Internet and how they are designed. For example, the technical design of the Internet's domain name system (DNS) has constructed or enabled certain forms of governance both of the DNS and by the DNS. As Bradshaw and DeNardis (2016) suggest, among other design features, the DNS embeds names (in the form of domain names) and therefore involves speech conflicts, its hierarchical design creates chokepoints at which content can be blocked, it involves a pool of finite resources (binary Internet addresses) and so raises issues of global distribution, and the requirement for unique names and numbers has necessitated centralized administration to ensure global uniqueness. Much prominent research has focused on the DNS and the systems of institutional control (e.g., the Internet Corporation for Assigned Names and Numbers [ICANN], the Internet Assigned Numbers Authority [IANA], registries, registrars) around this system (e.g., Klein 2002; Kleinwächter 2000; Mueller 2002; Paré 2003).

A significant Internet governance enterprise is the establishment of technical standards for universal formats for how to address, encode, compress, encrypt, and exchange information in a way that is interoperable with other devices and software that adhere to these standards. There are hundreds upon hundreds of core standards, but some of the most well-known are Wi-Fi, HTTPS (hypertext transfer protocol secure), VoIP (voice over Internet protocol), and Bluetooth. These specifications are set by many transnational technical institutions, such as the Institute of Electrical and Electronics Engineers (IEEE), the World Wide Web Consortium (W3C), and the Internet Engineering Task Force (IETF). Standards serve a technical function, but their design also establishes public policy. For example, interoperability standards allow economic competition among private actors and promote innovation and global interconnection. The strength of encryption standards establishes conditions for privacy. Web accessibility standards allow those with hearing, sight, movement, or other impairments to use the Internet.

The technological affordances of information intermediary technologies also enact governance. Information intermediaries are the private platforms (e.g., social media platforms, search engines, messaging platforms, access providers, and cloud computing companies) that enable the exchange and aggregation of content. Design features such as tracking mechanisms, real-name identification requirements, and decisions about anonymity construct rights in the same way that the terms of service of these systems construct the conditions of individual civil liberties such as speech rights, privacy, and data ownership.

Cross-Border Technologies and Bordered Policy Often Conflict

Policy making that seeks to stay within national borders does not necessarily stay within national borders. For example, the European Union's General Data Protection Regulation (GDPR) has affected companies that do business all over the world and has affected cross-border technologies such as the WHOIS system, a database of global registered domain holders. Local policies, such as the GDPR, the right to be forgotten rule, and data-localization policies, have global effects because the underlying technologies of the Internet—interconnection points, cloud computing infrastructures, content distribution networks, the DNS—do not correspond to the borders of these local policies.

Yet one political reaction to the growth and success of the Internet, the technological reality of cross-border technologies, and the market reality of private companies shaping human rights online is the rise of cyber sovereignty approaches in which countries are seeking to exert greater control over the Internet. In many cases, they are seeking to impose nation-state borders over the distributed architecture of the Internet. Russia, China, and some other countries have been proponents of cyber sovereignty under the guise of social order, with China's efficient system of content censorship and filtering perhaps the best example.

A long-standing debate in Internet governance involves those advocating for greater government control of the Internet, such as the cyber sovereignty models, and those advocating for the preservation of technical governance that is distributed over actors including international organizations, traditional governance structures, the private sector, new global institutions, and civil society. This has led to many international governance controversies, from debates over international telecommunications regulations at the International Telecommunication Union's World Conference on International Telecommunications in Dubai in 2012 to the controversial and long-coming transition of power from the US Commerce Department to ICANN in overseeing the IANA functions.

While the Internet has always been subject to national statutory contexts, its distributed architecture and other technical features make implementing individual national laws difficult in practice. Transnational private companies have to deal with unique legal requirements in all the markets in which they operate or even simply where users might access their services. Governments are also increasingly establishing policies that place constraints on technical infrastructure arrangements. For example, relatively new data-localization laws are in place from Russia to Latin America to Asia that place restrictions on how customer information is stored, often requiring data to reside on servers within a country's borders. These policies affect not only traditionally tech companies but any company (e.g., financial services, retail) that stores customer data. Some of these policies arose over concern about citizen privacy and foreign surveillance, but they create complications from an engineering, human rights, and business model standpoint. Concentrating data in one place can actually make it more difficult to protect personal privacy. These requirements also do not map onto the distributed technical design of the Internet or the ways in which content delivery networks (CDNs) decentralize and distribute data around the world. This is one example of the tension between national governance contexts and the distributed nature of the Internet.

The Privatization of Governance and the Multistakeholder Model

The term "Internet governance" is, in some ways, an oxymoron. The power to control and govern the Internet is distributed among private industry, global institutions like ICANN and the IETF, and in some cases by civil society, as well as by governments. This form of distributed governance is often called multistakeholder governance and involves functions such as technical-architecture-based enforcement of intellectual property rights, the private policies of technical intermediaries, the administration of the DNS and Internet names and numbers (i.e., IP addresses), cybersecurity coordination, and the establishment of Internet standards. Collectively, these tasks keep the Internet operational and enact policies that directly establish the conditions of innovation and civil liberties in the digital sphere.

A major line of inquiry in Internet governance studies involves questions about the nature and legitimacy of multistakeholder governance arrangements and the appropriate balance of powers among actors at the various layers of Internet coordination. Frankly, it is not yet a well-understood or much-analyzed framework of governance. There are many models and many contexts of multistakeholder governance. Drawing from John Ruggie's pioneering study of multilateralism, Raymond and DeNardis (2015) offer a taxonomy of different types of multistakeholder institutional forms that vary according to what combination of actor class is participating and the nature of the authority relations among these actors. The international relations scholar Joseph Nye Jr. describes Internet governance as a "regime complex," applying regime theory to Internet governance to explain the constellation of institutions, actors, norms, and policies that collectively constitute distributed, multistakeholder governance. As Nye explains, Internet governance involves "a set of loosely coupled norms and institutions that ranks somewhere between an integrated institution that imposes regulation through hierarchical rules, and highly fragmented practices and institutions with no identifiable core and non-existent linkages" (Nye 2014, 9).

A critical point for understanding Internet governance is that there is not a single system of oversight and coordination but an entire constellation of functions, each overseen by different governance structures distributed over one or more actors. Collectively, this administration and coordination of the technologies necessary to keep the Internet operational and the heterogeneous policies enacted around these technologies is viewed as distributed, multistakeholder governance, even if in practice multistakeholder arrangements rarely match the rhetoric around multistakeholderism. Thus, scholars study the policy-making role of private industry (DeNardis 2014; Gillespie 2014; MacKinnon 2011), national and international law (Goldsmith and Wu 2006; Weber 2010), technical coordination institutions (Klein 2002; Kleinwächter 2000; Mathiason 2008; Mueller 2002), international organizations (Levinson and Marzouki 2015), and as mentioned in the previous section, technical design itself (Braman 2011; DeNardis 2009).

Internet Security Is Converging and Diverging with National Security
There has long been a peculiar rhetorical distinction between "cyber" and "Internet" for historical and cultural reasons beyond the scope of this book. A person using the term "cyber" often refers to cybersecurity or national security domains of Internet warfare and international relations. Using "Internet" refers to either digital economy issues, the free and open Internet, or the Internet of Things (IoT).

From an engineering perspective, the distinction makes no sense, because the underlying infrastructure is the same, and when using "Internet governance" this book does not imply a distinction between cyber issues and Internet issues. They are the same.

It may be helpful, however, to acknowledge the ways in which the discourses and communities of practice around these nomenclatures are both converging and diverging, because this helps illustrate another feature of

Internet governance: that points of control are points of mediation between often-conflicting values.

Internet security and national security, on one hand, are converging because the stability of the economy, democracy, and public sphere is now completely predicated on Internet stability and security. Every sector of the global economy is digitally mediated and in some way connected to the public Internet. Security breaches have significant effects on basic societal functioning. Ransomware attacks have cryptographically locked and therefore crippled health care systems until the affected institutions succumb to paying ransom, usually in the form of Bitcoin. High-profile, massive consumer data breaches such as those at Equifax, Target, and the Office of Personnel Management have chilling effects on citizen trust in the digital economy and sometimes chilling effects on speech and behavior online. Even more significant, the proliferation of the IoT raises the stakes for security because an outage or disruption of cyber-physical systems can mean the loss of life and not just loss of access to communications. Stable systems of democracy also increasingly require strong cybersecurity, considering the stunning disclosure by US intelligence agencies about Russian probing of voter rolls and other cyber incursions during the 2016 US presidential election.

On the other hand, cybersecurity and national security are diverging. Other security trends, such as governmental stockpiling of zero-day exploits and the rise of cyber offensive capabilities such as the Stuxnet code targeting Iranian nuclear reactors, speak to cyber as the fifth domain of warfare and the emerging front for conflict between nation-states. The need for strong security for the digital economy and for individual privacy and trust in cyberspace comes into conflict with national security requirements for law enforcement, intelligence gathering, and the amassing of cyber offense capability. The clash of market-driven trends toward greater encryption with law enforcement requirements for access to data materialized in the aftermath of the San Bernardino, California, terrorist attack when authorities sought access to an encrypted Apple smartphone belonging to the attacker. Values are always in tension around Internet control points.

Control of Internet Governance Infrastructure Is a Proxy for Political Power

Global conflicts over control of cyberspace have existed at least since the commercialization and internationalization of the Internet. One prominent example was the long-standing geopolitical contention over the US Commerce Department oversight of names and numbers, including its contractual arrangement with ICANN and authority over changes to the root zone file, until this unique US coordinating function was transitioned to the global multistakeholder community. As the Internet's importance to the economy and the political sphere has increased, so has contention over the infrastructure of the Internet.

Governments and other forces recognize that power over technical infrastructure points of control can serve as a proxy for control of ideas, the economy, and the political sphere (DeNardis 2012; Musiani et al. 2016). For example, the DNS has become a tool of content control—for example, used by China's extensive censorship system and used to block access to sites that illegally share pirated content or sell counterfeit products. Encryption standards and implementations, historically always politically charged, have increasingly become targets of governments wishing to weaken or create backdoors to cryptography for national security or intelligence purposes, in some cases pitting law enforcement values against the need to provide strong security for the digital economy. As with all areas of Internet governance, battles over control of infrastructure are sites of conflict among competing values and interests.

The Ensuing Challenge to Internet Governance Scholarship

The preceding five themes in Internet governance practice translate directly into challenges, ab initio, for Internet governance research. A goal of this book is to demonstrate how to overcome clear research challenges in studying Internet governance.

Making the invisible visible. The technical architectures and institutions of governance are not visible in the same way that Internet content and usage is visible to end users. Scholarship has to excavate and make visible these hidden infrastructures, in some cases, before research even commences.

Understanding complex technologies. Studying the design and governance of the Internet requires understanding the underlying technologies. Technologies of Internet governance include thousands of protocols, platforms, algorithms, systems of routing and interconnection, the DNS, encryption standards, the Internet of things, and public key cryptography and other authentication mechanisms. These systems constitute the underlying infrastructure supporting both cyberspace and the integrated cyber-physical world. The application of machine learning and artificial intelligence as mechanisms of governance further complicates the topic of study. This complexity of technologies either enacting governance or being governed requires scholars, regardless of discipline, to have a proficient technical understanding of how stuff works.

The difficulty in studying the private sector. The private sector owns and operates the vast majority of the Internet's infrastructure and platforms, further complicating access to data and sometimes concealing technology in proprietary enclosure such as algorithms protected by trade secrecy or standards-based patents.

Navigating conflicting values. Because Internet governance points of control are increasingly political points of control, scholarship about these conflicts often takes on a normative stance. Even the choice of what to study in Internet governance intervenes. Almost every question of Internet governance embeds conflicting values, such as law enforcement versus individual civil liberties, privacy versus free speech, technical expediency versus security, surveillance capitalism versus privacy, and consumer safety in the IoT versus economic competition. Is a universal and interoperable Internet desirable or does a fragmented Internet that, for example, isolates industry-specific IoT applications have advantages? Even if objectivity in research is possible, Internet governance research, especially considering the high stakes to society, often involves some type of a normative stance, such as the assumption that a free and open Internet is desirable.

Multistakeholder governance and multistakeholder research. Because of how technology crosses borders and because even local governance decisions can have global effects, studying any one actor or issue area can sometimes miss important contextual or empirical factors. At the same time, collaborative research initiatives that combine input from actor classes have promise for tackling very large issue areas.

Overstudying open systems. The traditions of Internet governance in practice have been generally open in that dominant coordinating institutions like the IETF and ICANN allow participant observation and have made proceedings and mailing lists generally accessible relative to other more insular institutions. Because of the availability of more data, these institutions and underlying systems are asymmetrically overstudied relative to systems involving greater proprietary enclosure. The institutions that are more closed, including consortia and institutions in many emerging areas of technology, are very difficult to study.

Research often involves creation of technological tools. Because of the massive size and complexity of Internet infrastructure, as well as the need to sometimes digitally reach across borders to gather data, research often involves software mediation and the coproduction of technological tools. This is especially the case for studies that assess politically motivated outages and cybersecurity incursions but also for studies of how traffic flows through interconnection points and for large-scale network analysis of all kinds.

Who Studies Internet Governance and How Do They Study It?

Internet governance researchers excavate and examine the invisible Internet control points and the social, economic, and political implications of these control points. Internet governance research, commensurate with Internet governance itself, is hardly a monolithic practice but, rather, made up of discipline-independent but interacting fields as well as intrinsically interdisciplinary fields such as science and technology studies and communication studies.

The methodological approaches and tools are diverse: large-scale text analysis, network analysis, traditional statistical methods, discourse analysis, participant observation, interviews, and ethnomethodologies of all kinds. Even among this diversity, there is clearly an epistemic community of interdisciplinary scholars who have studied dimensions of Internet governance for decades but perhaps most visibly coalescing with the founding of the GigaNet in 2006 just before the inaugural United Nations Internet Governance Forum in Athens. In other words, scholars from law, economics, history, political science, science and technology studies, sociology, and beyond self-reflexively situate what they are doing as Internet governance research.

Because of the technical complexity of systems of Internet architecture governance, many of these scholars have a strong background knowledge in computer science, engineering, information technology, and specific expert knowledge about the Internet's underlying technical architecture. Indeed, this technical expertise, even when one studies laws (about technology), institutions (that design and administer technology), or private ordering (the companies that own and operate the Internet).

Research is often policy engaged or policy adjacent. Internet governance researchers are interested in real-world problems and the opportunity for creating an evidence base for policy decisions. Not surprisingly, policy initiatives have directly engaged researchers and commissioned work on specific topics. For example, the Global Commission on Internet Governance, a two-year initiative chaired by Carl Bildt, a former prime minister of Sweden, included the global, interdisciplinary Research Advisory Network that produced more than 50 original research papers in six research volumes on cybersecurity, fragmentation, access and interconnection, and more (Global Commission on Internet Governance 2016–2017).

A related feature of this research is that it sometimes overlaps with the practice of Internet governance. Scholars have been actively involved as participants in ICANN working groups, contributing to standards-setting initiatives, moving between higher education and policy appointments, serving as advisors to policy makers, and sometimes serving as consultants to industry. This translational and pragmatic role of some scholars is similar to scholarly engagement in other topical disciplines that engage in great problems in contemporary society.

One could divide Internet governance research in many ways—disciplinary approach, topical area of study, research methodology, or thematic or conceptual framework. The editors of this volume choose to highlight some disciplinary perspectives on studying Internet governance. Although a wide variety of fields—from computer science to political science to science and technology studies—are included in this book, it is of course a single volume and thus not sufficiently inclusive of all disciplines. The authors were selected according to who could provide a diversity of perspectives and are influential Internet governance thought leaders in their respective fields.

The book commences with historically grounded chapters by two experts in Internet governance and communication policy. In chapter 2, "The Irony of Internet Governance Research: Metagovernance as Context," the information policy expert Sandra Braman sets the stage by broadly defining Internet governance as including "not only efforts to regulate institutional, communal, and individual practices, content, and uses by geopolitically recognized governments but also decision-making and efforts with regulatory effects by private sector entities, whether those that have a legal status (such as third-party intermediaries that have legal identities as corporations) or those that do not (such as autonomous networks)."

Her chapter then situates questions of Internet governance in the longer trajectory of network regulation and socio-technical governance and explains how Internet governance entanglements are challenging concepts such as liability, governance, the rule of law, and the state. In chapter 3, "Inventing Internet Governance: The Historical Trajectory of the Phenomenon and the Field," Milton Mueller and Farzaneh Badiei examine the emergence and trajectory of Internet governance as a label, as an area of scholarly study, and as a real-world policy arena and explore the interplay among these spheres.

To policy makers and some scholars, Internet governance is too often exclusively understood through institutional lenses—governments, private companies, systems of politics, international organizations, ICANN, the IETF, and so on—with less attention to the agency and affordances of technology itself. The Internet at its core is a collection of technologies—protocols, routing and addressing infrastructures; physical equipment like fiber-optic cable, antennas, switches, and interconnection sites; algorithm-driven platforms; the DNS; applications; code; firewalls; encryption; and the like. Arrangements of technology are also arrangements of public policy. Not surprisingly, then, the field of science and technology studies (STS) has been influential in examining and making visible the reciprocal relationship between, on one hand, technologies of Internet governance and architecture and, on the other, society and the economy. Francesca Musiani explains the contributions and perspective of STS in chapter 4, "Science and Technology Studies Approaches to Internet Governance: Controversies and Infrastructures as Internet Politics." The chapter pays particular attention to how studies of controversies contribute to understandings of Internet governance. Most notably, and speaking to the urgent need to look beyond institutional frames, Musiani explains how STS perspectives— and especially approaches to infrastructure studies—examine the agency of nonhuman actors and the mediating governance role of infrastructure.

Some of the early US legal writing about Internet policy, such as Lawrence Lessig's influential book *Code and Other Laws of Cyberspace* (1999), conceptually followed STS themes, especially highlighting the ways in which technical architecture (as well as law, norms, and markets) shapes and constrains society. As the Internet became commercialized and globalized, legal scholars have continued to produce important Internet governance work on every imaginable subtopic of Internet governance, whether trademark concerns in the DNS, intellectual property rights protection online, or privacy laws. One of the challenges inherent in legal studies of Internet governance is that Internet technologies and institutions do not neatly reside within national borders. In chapter 5, "A Legal Lens into Internet Governance," legal scholar Rolf H. Weber helps explain the challenges and the ensuing considerations of legal harmonization, legitimacy, and multistakeholder legitimacy in spheres of Internet governance.

The perspectives of computer scientists have contributed greatly to understandings of how the Internet is controlled and what is at stake. This book includes a chapter from a prominent team of computer science researchers from the Web Science Institute at the University of Southampton: Wendy Hall, Aastha Madaan, and Kieron O'Hara. In chapter 6, "Web Observatories: Gathering Data for Internet Governance," they take up the question of the study of governance over the flow of data and content in the web ecosystem. They discuss the challenges of developing and re-creating methods for ethical and secure data gathering and sharing, and they propose an architecture for doing so.

Policy makers rely (or should rely) on analysis of empirical data in all areas of Internet policy, including cybersecurity, but sometimes these data (and analyses) can be inadequate to reliably inform policy-making decisions. Quantitative political science research sheds light on Internet governance trends and problems. One challenge in this arena is that the thing being studied continually expands and changes, making examinations intrinsically multivariable and also difficult to replicate as data change. The political science professor Eric Jardine takes up these challenges in chapter 7, "Taking the Growth of the Internet Seriously When Measuring Cybersecurity." Jardine addresses the lack of statistical normalization and other challenges such as the failure to control for "lurking confounders."

Another question in Internet governance research is how to study the private sector, which owns and operates the majority of cyber infrastructure and establishes policies through design of systems, terms of service, and institutional decisions that, in effect, govern. These private intermediaries include social media platforms, Internet service providers, content distribution networks, cloud computing providers, private DNS resolution providers, and many other categories of industry. An entire generation of doctoral student researchers across various disciplines is interested in studying this privatization of Internet governance instantiated in the decisions of intermediaries. Carrying out research projects about the privatization of governance is challenging because it requires access to often closed and possibly even trade-secrecy-protected data.

The methods for studying the surface area of content are well established, but there is much more research fluidity and difficulty in studying what is beneath content, the hidden mechanisms and sinews of power controlling the flow of content and establishing conditions for human rights and innovation. For example, interviews with leading thinkers from the private sector sometimes are governed by nondisclosure agreements, underlying algorithms and other control mechanisms are protected by trade secrecy laws, and infrastructures are increasingly shrouded in proprietary enclosure. The Danish human rights researcher Rikke Frank Jørgensen takes up the question of studying Internet governance by private intermediaries in chapter 8, "Researching Technology Elites: Lessons Learned from Data Collection at Google and Facebook."

Studying content itself, however, is a critical area of Internet governance research and one that involves enormous data stores: media coverage of Internet governance topics, terms of service, deliberations about Internet governance, and so on. There are enormous quantities of text that can help elucidate Internet governance problems, understandings and misunderstandings, and solutions. Thus, research that uses text mining contributes greatly to examining the Internet governance ecosystem. Derrick Cogburn addresses this topic in chapter 9, "Big Data Analytics and Text Mining in Internet Governance Research." He studied 12 years of transcripts of the UN Internet Governance Forum to illuminate core themes and issues over time and determine the utility of text mining and big data analytics in Internet governance research on all topics, from censorship to cybersecurity.

The sheer volume of deliberations and discussions that feed into decisions about the design and administration of Internet architecture is massive. Because of the traditions of openness, transparency, and participation in Internet design communities, much (but not all) of this deliberation happens in the open and is archived in mailing lists, meeting minutes, and other online repositories. The Internet governance scholars Niels ten Oever, Stefania Milan, and Davide Beraldo address the topic of mailing-list research in chapter 10, "Studying Discourse in Internet Governance through Mailing-List Analysis." The authors explain the utility of and opportunity for interrogating mailing-list archives and propose a mixed-methods approach, a hybrid of computational and interpretive tasks.

Arguably the most societally consequential area of inquiry around governance of the Internet is cybersecurity. The global economy is completely digitally mediated and therefore dependent on the security and stability of networks. Privacy requires strong encryption. Consumer safety now depends on security of the IoT. Democracy requires not only secure voting systems but secure voter rolls and email. Free speech requires circumvention tools that provide freedom from filtering and censorship. Web queries require public key encryption. Online transactions require strong authentication. The study of cybersecurity governance is possibly the most critical area of Internet governance research because every other area depends on the security and stability of networks. The leading Internet governance researcher in this area is probably Citizen Lab director Ron Deibert, whose work has been groundbreaking because it has required the development of new technical tools as part of researching Internet governance. In chapter 11, "The Biases of Information Security Research," Deibert raises a critical point: even a highly technical area such as cybersecurity is politically contested and shaped by a constellation of economic and social factors that construct what research gets done. Chapter 11 raises questions of epistemology as much as methodology.

Dominant Internet governance discourses, as they have been constructed by those with a stake in the outcome of many tangible policy debates, have ideologies. For example, China and Russia have increasingly espoused an ideology of cyber sovereignty that advocates for strong nation-state control of the Internet in the name of order and as a reflection of authoritarian tendencies toward information and communication technologies. In the West, the collective coordinating tasks that keep the Internet operational have been cast, with some descriptive accuracy, as private-sector-led, multistakeholder Internet governance. But in the same way that cyber sovereignty embeds an ideology and privileges an approach, multistakeholder governance advocacy has been adopted as a way, for some, to oversimplify Internet governance as a monolithic practice or to preserve hegemonic power for dominant institutions.

In chapter 12, "The Multistakeholder Concept as Narrative: A Discourse Analytical Approach," a leading Internet governance scholar, Jeanette Hofmann, examines how narratives and imaginaries, including those emanating from academic research, are significant constructors of policy discourses. As Hofmann suggests, there is often a disconnect between expectations of multistakeholderism and how this model performs in practice. She explains how discursive representations of such concepts in Internet governance, including those coconstructed by the academic community, take on various utilities. The chapter addresses a critical subject but also helps emphasize the place of discourse analysis in Internet governance scholarship.

The book concludes with chapter 13, "Toward Future Internet Governance Research and Methods: Internet Governance Learning," in which Nanette Levinson draws from related research arenas to elucidate what foundations can inform the future of Internet governance research and methods.

Both "Internet" and "governance" are malleable terms whose meanings are in flux, especially as it becomes more and more difficult to define what the Internet is, whether based on underlying technical architecture, user communities, or underlying values. More "users" are bots and things than people. More networks increasingly depend on proprietary protocols, especially in cyber-physical systems, rather than open protocols such as TCP/IP (transmission-control protocol/Internet protocol). The Internet in China bears no resemblance to the Internet in Sweden. Acknowledging the spectrum of technologies, the conflicts between values, and the fragmentation of the Internet that already exists does not negate the descriptive reality of the present moment. The constellation of governance and control issues around the Internet now determines conditions of privacy, speech, innovation, and the security and stability of the digital economy. Internet governance researchers seek to shed light on these critical decision points that will shape society for an entire generation.

References

Bradshaw, S., & DeNardis, L. (2016). The politicization of the Internet's domain name system: Implications for Internet security, universality, and freedom. *New Media & Society, 20*(1), 332–350.

Braman, S. (2011). The framing years: Policy fundamentals in the Internet design process, 1969–1979. *The Information Society, 27*, 295–310.

DeNardis, L. (2009). *Protocol politics: The globalization of Internet governance*. Cambridge, MA: MIT Press.

DeNardis, L. (2012). Hidden levers of Internet control: An infrastructure-based theory of Internet governance. *Information, Communication & Society, 15*(5), 720–738.

DeNardis, L. (2014). *The global war for Internet governance*. New Haven, CT: Yale University Press.

DeNardis, L., & Musiani, F. (2016). Governance by infrastructure. In F. Musiani, D. Cogburn, L. DeNardis, & N. S. Levinson (Eds.), *The turn to infrastructure in Internet governance*. New York, NY: Palgrave MacMillan.

Gillespie, T. (2014). The relevance of algorithms. In T. Gillespie, P. Boczkowski, & K. Foot (Eds.), *Media technologies: Essays on communication, materiality, and society* (pp. 167–194). Cambridge, MA: MIT Press.

Global Commission on Internet Governance. (2016, December 7). *A universal Internet in a bordered world: Research on fragmentation, openness and interoperability* (Research Vol. 1). Retrieved from https://www.cigionline.org/publications/universal-internet -bordered-world-research-fragmentation-openness-and-interoperability

Global Commission on Internet Governance. (2017, January 7). *Who runs the Internet? The global multi-stakeholder model of Internet governance* (Research Vol. 2). Retrieved from https://www.cigionline.org/publications/who-runs-internet-global -multi-stakeholder-model-internet-governance

Global Commission on Internet Governance. (2017, May 8). *Mapping the digital frontiers of trade and intellectual property* (Research Vol. 3). Retrieved from https:// www.cigionline.org/publications/mapping-digital-frontiers-trade-and-intellectual -property

Global Commission on Internet Governance. (2017, June 23). *Designing digital freedom: A human rights agenda for Internet governance* (Research Vol. 4). Retrieved from https://www.cigionline.org/publications/designing-digital-freedom-human-rights -agenda-internet-governance

Global Commission on Internet Governance. (2017, July 26). *Cybersecurity in a volatile world* (Research Vol. 5). Retrieved from https://www.cigionline.org/publications /cyber-security-volatile-world

Global Commission on Internet Governance. (2017, July 26). *The shifting geopolitics of Internet access: From broadband and net neutrality to zero-rating* (Research Vol. 6). Retrieved from https://www.cigionline.org/publications/shifting-geopolitics-internet -access-broadband-and-net-neutrality-zero-rating

Goldsmith, J., & Wu, T. (2006). *Who controls the Internet? Illusions of a borderless world*. Oxford, UK: Oxford University Press.

Klein, H. (2002). ICANN and Internet governance: Leveraging technical coordination to realize global public policy. *The Information Society, 18*(3), 193–207.

Kleinwächter, W. (2000). ICANN between technical mandate and political challenges. *Telecommunications Policy, 24*(6–7), 553–563.

Lessig, L. (1999). *Code and other laws of cyberspace*. New York, NY: Basic Books.

Levinson, N. S., & Marzouki, M. (2015). IOs and global Internet governance inter-organizational architecture. In F. Musiani, D. Cogburn, L. DeNardis, & N. S. Levinson (Eds.), *The turn to infrastructure in Internet governance* (pp. 47–72). New York, NY: Palgrave MacMillan.

MacKinnon, R. (2011). *Consent of the networked: The world-wide struggle for Internet freedom*. New York, NY: Basic Books.

Markham, A., & Baym, N. (Eds.) (2008). *Internet inquiry: Conversations about method*. Thousand Oaks, CA: Sage.

Mathiason, J. (2008). *Internet governance: The new frontier of global institutions*. New York, NY: Routledge.

Mueller, M. (2002). *Ruling the root: Internet governance and the taming of cyberspace*. Cambridge, MA: MIT Press.

Musiani, F., Cogburn, D., DeNardis, L., & Levinson, N. S. (Eds.). (2016). *The turn to infrastructure in Internet governance*. New York, NY: Palgrave MacMillan.

Nye, J. S. 2014. *The regime complex for managing global cyber activities*. Global Commission on Internet Governance Paper Series (Paper no. 1). Centre for International Governance Innovation/Chatham House.

Paré, D. (2003). *Internet governance in transition*. Lanham, MD: Rowman & Littlefield.

Raymond, M., & DeNardis, L. (2015). Multistakeholderism: Anatomy of an inchoate global institution. *International Theory, 7*(3), 572–616.

Weber, R. H. (2010). *Shaping Internet governance: Regulatory challenges*. Berlin, Germany: Springer.

2 The Irony of Internet Governance Research: Metagovernance as Context

Sandra Braman

Even though reality
may not exist,
we have a right to it.[1]

Information policy—laws and regulations pertaining to any aspect of information creation, processing, flows, and use or, more colloquially, law and policy for information, communication, and culture—matters because it creates the context within which all other decision-making takes place. Internet governance, a form of information policy, matters in particular: it provides the context for much of that context, as the Internet is a "pan-medium" (Theall 1999), infrastructure for all forms of communication previously mediated by distinct technologies for which, historically, laws and regulations were differentially developed, interpreted, and applied. Critically, "all other decision-making" includes the processes of governance themselves.

At the close of the second decade of the 21st century, when profound challenges to rule of law are underway around the world, Internet governance researchers must grapple with the effects of the uses of this socio-technical system—and of decisions about what those uses might be, how they might proceed, and what the consequences are likely to be—on governance itself. This chapter situates Internet governance research relative to the nature of governance and metagovernance more broadly, taking steps toward a research agenda by identifying questions raised by these developments. Additional theoretical and conceptual work is needed to provide a foundation for analysis of dimensions not historically considered but fundamental to arguments and operations in a world of algorithmic agency and digital structure.

The chapter concludes with a few thoughts on humility, which brings us to irony; in Samuel Beckett's (1953/2012) words, "I can't go on, I'll go on." Bob Jessop (2016) uses the term "irony" to refer to analyzing, making, and implementing policy in the face of knowledge that, ultimately, all governance efforts will fail. For those who study network policy, his use of the concept appropriately resonates with Robert Britt Horwitz's *The Irony of Regulatory Reform* (1989), a model of network policy analysis that is fully imbued with the role of networks as agents as well as subjects of power, deeply involved in transformations of the state.

The Internet and Governance

Relationships between the Internet and governance go both ways. As DeNardis and Musiani (2016) succinctly put it, there is governance *of* the Internet, and there is governance *by* the network. The latter includes structurational and constitutive effects of Internet design and policy, whether direct and evident or indirect and needing analysis to become visible, as well as uses of the Internet as policy tools. The same elements can serve both "governance of" and "governance by" functions (Merrill 2016).

Many of the questions those involved in Internet governance engage are not new. Some are. It can take deep knowledge of history to know which questions are which; neither legal nor discursive silos help. In one example of their cost: the first edition of the *Tallinn Manual* (Schmitt 2013), the NATO-sponsored effort by international legal experts to determine whether and how existing international law pertains to cybersecurity and cyberwarfare, ignored network-specific treaties altogether. By the time of the second volume, *Tallinn Manual 2.0* (Schmitt 2017), the experts involved had apparently been exposed to Anthony Rutkowski's (2011) history of the treatment of what we now call cybersecurity issues beginning with the first international telecommunications treaties in the mid-19th century. Including this domain of international law in the *Tallinn 2.0* analysis affected a number of conclusions reached, but that came several years after the first volume had been released to inform other national and international decision-making.

Whether the questions Internet governance researchers face are new and unique to the context or not, the conditions of the world for which decisions are being made are qualitatively new in many ways, some of which we are only now beginning to discover and others of which are yet to come or may remain indiscernible. Even before the turn toward political extremisms of the first decades of the 21st century, the conditions under which we operate, and how we operate, had been changing in often radical ways and becoming increasingly turbulent—so much so that political scientists feel driven to add the prefix "meta-" to the words they are using as they think about what is going on. Whether the conversation starts with what is happening to the state (government), or whether it starts from what governs in a particular area of social life (governance), it leads to metagovernance. Indeed, as Meuleman (2008) demonstrates, quite different intellectual traditions wind up in essentially the same place in this regard. Metagovernance involves establishing system parameters and determining what can happen within and between systems. Information policy, including Internet governance, is inherently parametric policy, a matter that Lawrence Tribe (1985) notes makes it constitutional in nature within US law and that Jessop (2011) identifies as so important that engagement with parametric issues is itself one of the five types of metagovernance in his handbook chapter on that concept. It includes material as well as normative structures, making Internet governance infrastructure in this additional way for all other forms of metagovernance and governance, too.

The keystone works on the state and governance of recent years referred to in this chapter were published before the 21st century ruptures in countries around the world. That does not lessen our responsibilities as Internet governance researchers. This section looks briefly at some of the theoretical literatures that have been or could be useful for thinking about Internet governance within the larger context of the evolution of forms of governance and metagovernance before turning to the regime theory that is important to several other chapters of this book—and at its limits.

Theoretical Context

This is a world in which theoretical pluralism is not only preferred but a necessity. Jessop (2016) presents several theoretical approaches to study of the state, all of which he argues have validity and importance. We should not expect the directions in which things evolve to be singular. At their extremes, as Marshall McLuhan notes, the effects of the use of information and communication technologies can simultaneously be opposite in nature (McLuhan and McLuhan 1992). Where there are singularities, rationalities can bifurcate (DeLanda 1991). The causal processes that get us from here to there are not necessarily linear, despite either desire or perception (Martine 1992). Where positivists would look for determinism, matters may well be stochastic. Policy concepts that have been useful in the social environment may not be so on the technical side, or may yield quite different outcomes (Edwards and Veale 2017).

The characteristics that make informational metatechnologies such as digital technologies qualitatively distinct from modern technologies and premodern tools (Braman 2002) have transformed the nature of the matériel through which system transformations take place (Archer 1982, 1984; Giddens 1984) to such an extent that whether something is agent or structure can often now be a matter of choice. This identifies an Internet governance research agenda of its own. We have long been aware of the agency/structure choice when it comes to intellectual property rights, a domain in which manufacturers can often put the same capacities into either software or hardware, respectively relying on either copyright or patent for protection. David D. Clark (2018) identifies and explains other significant areas in which the option is available in ways that have governance implications as we work on the "future Internet."

McKelvey's (2018) work on daemons, a concept important in technical design of what we now call the Internet from the earliest years of that process (Braman 2011), draws our attention to network potentialities that are unseen and unknown for most users, latent until they become active. This gives us a user-, rather than network-, oriented approach to thinking about the technical environment and what it is like to live as a human within a world of algorithms. The development of positive policy recommendations for a daemon-filled world of the type described by McKelvey leads in the direction of capabilities, so persuasively introduced to information policy by Julie Cohen (2012). There is work to be done here.

Regime Theory

The regime theory relied upon by a number of authors in this book was developed by political scientists as a way of thinking about international relations in issue areas where things were generally working but for which there was no existing international law—and, when it came to digital network matters, often no existing national law, either. The rising salience of such issues was reflected by policy analysts, many of whom shifted their attention from government to governance during the 1990s (Bache et al.

2015). Working with the framework as classically presented by Krasner (1983), Cowhey (1990) was the first to explicitly use regime theory in analysis of telecommunications policy. Mueller's (2004) work comparing the Internet and satellite governance mechanisms provides another example in network policy terrain. It is useful in analysis of a number of specific information policy issues raised within the context of the Internet as well (Braman 2004a, 2004b). Drezner (2009) and Nye (2014) are frequently referred to by those who explicitly apply this approach to Internet governance. Although regime theory was initially conceived of in regard to international relations, it has been taken up for use at other levels of social and governance structures, all the way down to the municipal level (Coletta and Kitchin 2017)—also matters of Internet governance because, as we know from Star and Ruhleder (1996), it is in its local manifestations that infrastructure comes into existence. It would be valuable to have a mapping of the multiple levels at which regime formation and transformation are now taking place in ways that are pertinent to—or comprise—Internet governance across all these levels.

A corollary of regime theory is that private sector agents are explicitly and openly—rather than begrudgingly, critically, and/or with hesitation—included within policy analysis. International law firms, hired to craft contracts for network-reliant and globally active clients, were already playing significant roles in thinking about legal arrangements for transnational digital information flows in the 1980s (see, e.g., Bruce, Cunard, and Director 1986), influencing public law affecting the Internet in inevitably path-dependent, if not precedential, ways. A related set of processes was underway in the network-intertwined industry of finance (Dezalay and Garth 1996). What would the kind of analysis undertaken by Dezalay and Garth on finance yield if undertaken on the role of such players in Internet governance?

As regime experience accumulates, working rules can become formal law, and norms can become foundational or operational principles. The kinds of learning Levinson talks about in her concluding chapter 13 of this book are among those involved. How do formal processes of Internet-specific institutions impede or encourage such learning? How useful is regime theory under conditions in which learning does not apply, whether because it is in a postlaw rather than prelaw context, or because rule of law has collapsed?

Regime theory involves governance under conditions of invention, evolution, or transition that can reify, whether as laws and regulations of geopolitically recognized states or through other means. Regime theorists typically assume incremental change in political and legal conditions, and that what are perceived to be very rare possibilities need not be taken into account. We now know, though, that change can be radical rather than incremental, sudden rather than slow, and that the outlier possibility may be the one we have to live with. Those doing Internet governance research need to be thinking about what happens under conditions of exception, of crisis, of turbulence, and of chaos. What role does Internet governance play in abuses of human rights and civil liberties? What roles might it play in preventing or mitigating such abuses? The burgeoning global community of researchers focused on resilience should be an important venue, and set of collaborators, for Internet governance researchers. The UN Sustainable Development Goals are now commonly being taken into account across issue areas; how might that be done with Internet governance research, following in the footsteps of Rajnish et al. (2017)?

Implications for Internet Governance Research

These elements of the theoretical context for Internet governance research make the terrain for this book's authors a set of "meta" questions. What kind of research do we need to help design infrastructure for imagining, desiring, and creating governance systems for the world that we irrevocably, and together, now inhabit? How can we understand effects of policy decisions that are stochastic and iterative in nature, much like financial derivatives, with declining degrees of confidence in predictability? How should we govern governing? The next section provides some foundational definitions for addressing such questions.

Definitional Basics

The concepts of the Internet and of Internet governance are addressed here. Discussion of the intervening concept of governance includes attention to metagovernance as well.

Internet

I join Abbate (2017) and Russell (2017) in the analysis that we are nearing the end of the period during which the concept of the Internet is the dominant frame for policy and/or many other purposes. (Because there are those who perceive themselves not to be on the Internet but on Facebook or some other platform, the extent to which this already affects perceptions of Internet governance matters among users. The ways in which they become involved in governance would be an interesting research question.) But I also agree with Hofmann, in chapter 12, and with Jessop (2016), in his work on the semantics of governance, that there are political power and policy efficacy in rhetorical frames and narratives. Thus I also take the position that thinking in terms of Internet governance will continue to have utility even as the actual network merges with material, biological, and social environments. That leaves us with two definitional problems—which legal history pertains, and how do we know when what is being regulated is communication and therefore fundamentally a matter of human rights and civil liberties?

The editors of this volume, like those of the journal *Internet Histories* (Brügger et al. 2017), are to be lauded for not only acknowledging but also actively encouraging appreciation of the multiplicities of the technological realities and experiences of the Internet. There are times, though, when which history is being privileged matters. It is this that is the foundational question in the network neutrality battle being played out in the United States. Ithiel de Sola Pool's seminal book *Technologies of Freedom* (1983)— which argues that, as different legal systems converge to cope with the convergence of computing and communication technologies, it is likely that the most repressive features of each would dominate in the new system that will emerge—explains how the trifold technological history (print, telecommunications, and broadcasting) yielded three different regulatory systems in the US by the time that government began to deal with digitization. Two of them, systems with very different regulatory approaches, have both been discussed as providing the history of the Internet for the purposes of network neutrality and other regulation. Additional legal histories are available when the Internet is approached from perspectives that are not oriented around communication. It would be the separate histories of currency, finance, and capital that apply to what has been described as "the internet of money" (Libra Association Members, n.d.), a suite of offerings that includes cryptocurrency and the associated financial services Facebook and a group of corporations that at launch included Mastercard, Visa, Spotify, Uber, and Lyft have been promoting. Starting from finance rather than communication frames Internet governance histories in ways that preference values oriented around capital rather than human rights or citizenship. Similar stories can be told for every other government around the world.

As Abbate (2017) points out, how Internet governance is defined is itself a political question precisely because it determines which histories pertain, "ideology in practice" (Gurumurthy and Chami 2016, 1). This is not just inherent in the design process, but is also a matter of deliberate effort: "Internet governance technologies not only embed political values in their design and operations but are increasingly being co-opted for political purposes irrelevant to their primary Internet governance function" (DeNardis 2012, 721). Which histories should provide the frame for Internet governance? Should different histories serve as foundations for diverse dimensions of the problem? Is it useful to incorporate historical pluralism as well as theoretical pluralism in Internet governance research?

Policy analyses in any specific area can often be greatly enriched by looking across diverse regulatory systems for the various ways in which the same type of problem has been addressed. One example: the issue of who controls the interface between private and public environments has arisen for both material and electronic networks, but the policy discussions for each have taken place within different legal silos and have not cross-referenced each other despite the shared features of the problem. On the material side, in the United States, treatment of mailboxes was a matter of constitutional law; on the electronic side, attachments to customer premises equipment are matters of administrative law, regulated by the Federal Communications Commission (FCC). Debates over the latter, which launched the liberalization of telecommunications regulation in the United States in the 1970s, were intense but never referred to the constitutional issues, although arguments made and principles used in constitutional law would have been pertinent in ways that have become ever more obvious.

The Internet border gateway protocol (BGP), critical to relations within and between the autonomous systems of which the Internet is made and thus key to human rights issues such as efforts to censor or shut down national networks altogether (Vargas-Leon 2016), is different in many ways from either the physical or electronic predecessor issues, but the discussions on these matters went on for a long time, arguments were presented from a wide range of perspectives, and there is a great deal that could be learned for the toolbox of concepts and possibilities to consider for Internet governance. Not every argument will transfer, but it is also likely that many will, as adapted or reoperationalized for the context. The point is not specific to the border gateway protocol.

Crawford (2007) prescriptively suggests that the Internet governance conversation should meld with the general domain of communication law and policy. If that is so, there is still the difficulty of deciding when Internet governance issues should be decided in light of fundamental constitutional principles and human rights law, justified when the network is understood to be about communication, arguably balanced differently against other needs when what is at stake is a transaction or a weapon. We are a couple of decades now into struggles over treating software as speech (see, e.g., Burk 1997; Coleman 2009) and are beginning to see the literature using speech-related arguments in analyses of autonomous entities such as robots (see, e.g., Bambauer 2017; Calo, Froomkin, and Kerr 2016). Quite aside from what speech rights would inhere to autonomous digital agents should they be granted any form of citizenship status, there is no obvious limit to the range of types of digital information collection, processing, and flow issues to which free speech analyses might be applied.

A look at the many and diverse ways in which US lawmakers and regulators have historically tried to bound the field of communication, published not long after the beginning of the 21st century, concluded that the most useful means of doing so for this technological era would be to treat media policy as those matters that mediate the nature of the public itself—who and what it is, the conditions under which members of the public can discuss together shared matters of public concern, and to which information members of the public have access upon which to base their discussions (Braman 2004c). That would not be a bad place to begin to think about how to draw boundaries regarding what should be considered to be communication for the purposes of Internet governance research, with the important caveat that it is now clear that several additional dimensions of analysis, discussed below, would need to be added to evaluate whether any given design decision, regulation, or content policy is now required.

Governance

A handbook by political scientists on governance opens by defining the concept in very general terms: "Theories and issues of social coordination and the nature of all patterns of rule" (Bevir 2011, 1). This is the broadest of possible approaches, including theories and practices that increasingly involve governance as hybrid and multijurisdictional, populated by a plurality of stakeholders who engage with each other via networks. A synthesis of the governance literature as it has developed to include metagovernance refers to it as the coordination of structures and practices that themselves are involved in coordinating social relations marked by complex, reciprocal interdependence (Jessop 2011). Metagovernance can be unbundled into first-order efforts involving one form of metagovernance, and second-order activities that involve multiple forms of metagovernance.

Law (1992, 382) memorably describes governance as "an effect generated by heterogeneous means." The comment is not just accurate and witty but also has an implication that is profound: *Governance is always emergent*, in the specific sense in which that concept is understood within complex adaptive systems theory—an emergent system is one that cannot be explained at any other than the system level, rather than by the operations of any of its parts. Epstein, Katzenbach, and Musiani (2016) go further, arguing that by definition all the specifics of governance can *not* be identified; that is, *governance is fully recognizable only post hoc*—by the time you see it, it is in place. (Josephine Wolff [2018] titled her book on the cybersecurity dimensions of Internet governance *You'll See This Message When It Is Too Late*.) This post hoc feature of visibility makes preemption, as understood theoretically by Brian Massumi (2007), particularly important. From this perspective, the comment from Karl Rove, aide to President George W. Bush, regarding the "reality-based community" may be not a cynical throwaway but an empirical description of importance to researchers, courts, physicians, and others oriented around the facts. As it was first published in *The New Yorker* by the reporter to whom the statement was given:

> The aide [Rove] said that guys like me [the reporter] were "in what we call the reality-based community," which he defined as people who "believe that solutions emerge from your judicious study of discernible reality."... "That's not the way the world really works anymore," he continued. "We're an empire now, and when we act, we create our own reality. And while you're studying that reality—judiciously, as you will—we'll act again, creating other new realities, which you can study too, and that's how things will sort out. We're history's actors... [ellipsis in original] and you, all of you, will be left to just study what we do." (Suskind 2004)

Jessop (2011) argues that governance inevitably fails, appearing to be successful only when dilemmas are framed so that negative, sometimes catastrophic, consequences are beyond the spatiotemporal horizons of visibility. Each of these categories of dilemma frames suggests a research program for those studying Internet governance. Corporation Schlumberger intentionally designed instruments and information collection practices for oil and gas exploration to be deceptive to governments, customers, and the public with respect to just which information was actually being collected, playing with such horizons of visibility for persuasive and operational purposes (Bowker 1994). What would such social technologies and practices look like in the domain of Internet governance? Over time, government structure and departmental or agency design can function as spandrels, analogous to architectural features that were once structurally necessary but that are now available for aesthetic, rhetorical, or other purposes (Braman 2006). Which features of Internet governance might be in use as spandrels, making it appear as if something presented is offered because it is considered desirable by users, the public, or policy makers when in reality it is something else, as well as or instead, that may for producers or others be the real point? In what ways can or does Internet governance make use of the spandrels of national, regional, and other governments? Are there equivalents to these questions regarding spandrels as they would apply to the corporate decision-making so important to Internet governance?

There has been so much scholarly discussion about changes in the nature of the state since the 1970s that it is actually identifiable as a distinct literature, on changing states. Three streams in this literature have arisen over time (Bevir and Rhodes 2011). The first engages the networked state as it has been practiced and understood by the late 1980s (see Antonelli 1988), a form in which the multiplicity of networked relations at every level of the governance structure creates an environment in which the state can no longer exercise power unilaterally. The second involves work on metagovernance, theories about ways in which the state continues to exert control in the networked environment by managing the multiple processes in play. The third loosened the sense of state control even further, focusing on "decentered governance," abandoning both governance and government in favor of attention to how individuals and elites exercise power in an environment framed more in terms of ethnically based nationalisms than the bureaucratic dimensions that typically characterize analyses of states. This sense of the decline in the effectiveness of governments has been expressed in various ways. Jessop (2004) discusses governance in the shadow of government. For Bovaird (2005), it is governance without government. For Bevir and Rhodes (2011), the stateless state. Arguments can be made that there are areas of Internet-based activity that may be beyond our ability to govern at all (Braman 2015).

Under these conditions, it is reasonable to ask whether continuing to engage in policy making and analysis is meaningful. Jessop suggests the concept of "collibration" to refer to a variety of techniques that governments use in situations in which there would otherwise be regulatory failure. He likes that concept because it works across types of policy tools and processes, but this is an area in which there has been a lot of creativity. The Organization for Economic Cooperation and Development (OECD 2018) offers a variant in its report on science, technology, and innovation, another area of information policy that Antonelli (2017) argues falls within the domain of Internet governance because of the network's importance to knowledge production and knowledge production's importance to society, the economy, and governance writ large. The OECD report uses the term "concertation" to similarly refer to efforts to hold things together that are so various in kind, so multiple in number, and so distributed across levels of the social structure, levels of the governmental hierarchy, and geography that effective management would in reality, most experienced practitioners and observers believe, be a rank impossibility. The performative value of theoretical work is being emphasized throughout this essay, but it has its limits. Asserting new concepts in any area is no guarantee it will make things happen. It is this recognition of the limits of one's efficacy in the face of expected failure that leads Jessop (2011) to insist that one principle for successful metagovernance should be "requisite irony," to which this chapter returns in its conclusion.

There is a substantial literature further articulating governance conceptually and analyzing it in various contexts to which references here and in other chapters in this book will point the reader. One takeaway from this work for those doing Internet governance research would be to abandon the sense of exceptionalism that continues to hover, even if more faintly than before, over the research community. Continuing to learn about other domains in the manner modeled by Raymond and DeNardis (2015) is not only valuable for analysis of these matters as they pertain to the Internet, but is the only way to fully understand the role that Internet governance plays in larger governance and metagovernance processes.

A second takeaway is the importance of acknowledging developments in the systems with which those involved in Internet governance are engaging. Easy assumptions cannot necessarily be made about which countries fall into which category when it comes to characterizing their political nature, for example. Jessop's authoritative work on theories of the state published in 2016 still refers to the United States as a liberal democracy, even though earlier work of his describes exactly the steps through which the political affairs of 2020 are unfolding with prescient clarity. There are extreme developments in many other countries around the world. Long-standing assumptions underlying political, legal, and policy analyses need to be unearthed and questioned or the analyses of Internet governance researchers, too, will be limited to historical matters.

Internet Governance

Musiani (2015) provides a valuable review of the literature on conceptualizations of Internet governance. This chapter is placed within a simplified typology of types of definitions that has a core shared across all types of definitions, the most *narrow* approach, applying only to management of the network itself. From this perspective, exemplified by Mueller and Badiei in chapter 3, Internet governance is the responsibility of those global institutions explicitly created for that policy purpose and devoted to it—ICANN, the Internet Engineering Task Force, the Internet Architecture Board, and related entities. An *intermediate* definition of Internet governance would add "uses" to the subjects of governance and the national institutions responsible for the Internet in their countries, such as the Russian Runet as imagined in the "Internet isolation" (*Moscow Times* 2019)—or "reliable Internet" (Tass 2019)—bill of 2019. The approach offered by DeNardis in chapter 1, building on her earlier work with Musiani, in my view relies on an intermediate definition. I use a *broad* definition of Internet governance that includes not only efforts to regulate institutional, communal, and individual practices, content, and uses by geopolitically recognized governments but also decision-making and efforts with regulatory effects by private sector entities, whether those that have a legal status (such as third-party intermediaries that have legal identities as corporations) or those that do not (such as autonomous networks).

There are two different ways of seeking the literature in a given policy area. One is to bound the domain through the lens of a specific term or set of terms (focusing on words used). Mueller and Badiei in chapter 3 use this approach to analyze the history of the Internet governance literature. Browne's (1997) approach to defining "information policy" does the same with that related and pertinent concept. At the opposite end of the spectrum, the domain can be bounded conceptually and theoretically, irrespective of the terminology used by various authors to refer to elements of the domain. My approach to defining information policy represents this end of the spectrum for that concept (Braman 2006), and a similarly broad approach to bounding the domain of Internet governance is used here.

Searching on the phrase "Internet governance" to locate literature on the subject will be most successful with the narrowest definition, though that still will not yield a comprehensive view of the pertinent literature because other terms continue to be used to refer to the same sets of institutions, functions, and activities. It will have some but less utility with an intermediate definition. Searching on that phrase will be least successful with the broadest definitional approach, missing a great deal in the pertinent literatures. Use of the narrowest approach is most valuable within academia (where resource battles begin by bounding turf, with its genuine implications for things such as faculty positions and budgets) and, of course, operationally for those involved in Internet-specific decision-making processes. Intermediate approaches can have enormous utility for analytical purposes. General public discourse about Internet-policy-related matters and the experience of individuals, communities, and organizations typically use the broadest approach, so in my view that can be particularly valuable for outward-facing communications of academics in addition to its heuristic and analytical value.

Thus, here "Internet governance" includes uses (and users), decision-making by general-purpose policy-making entities (e.g., geopolitical governments) and by those specific to the Internet (e.g., ICANN), and decision-making and structurational actions by private and public sector entities, daemons and humans, through informal and formal, transient and fixed, means. This is the same approach taken in a mid-1990s bibliographic essay on the streams of literature in areas of the law that were coming together into an identifiable field of Internet policy (Braman 1995), much as happened with the microeconomics of information and the macroeconomics of the information economy (Braman 2005), with two differences: today Internet-specific entities loom much larger, both in the pertinent literatures and in my thinking, and we are now using the frame of algorithms to address issues in this space as well.

What we think of as Internet governance issues will remain of central importance even when the term "Internet" itself has become a limited referent for the intelligent network environment within which we govern and are governed, and even when we are talking about metagovernance in all its versions rather than only governance. It has value because it keeps our attention on the range of existing decision-making venues from the global to the local, is a constant reminder to think about interactions among the effects of different decisions and policies, offers a singular lens onto complex interactions among many policy issues, and provides a valid umbrella for the range of types of decision-making venues, processes, stakeholders, and effects involved. Finally and importantly, thinking in terms of Internet governance justifies reliance on constitutional principles and international human rights law. It is not coincidental that David Kaye, the UN special rapporteur for freedom of expression, titled his 2019 book *Speech Police: The Global Struggle to Govern the Internet*. Rebecca MacKinnon used her 2012 book title to emphasize the importance of including Internet users in the network's governance if there is to be adequate concern for human rights and civil liberties: *Consent of the Networked: The Worldwide Struggle for Internet Freedom*.

Internet Governance and the State

Of the five trends in Internet governance Laura DeNardis discusses in chapter 1, two involve challenges to the state. The first, one of the most important arguments in that chapter, starts from the perspective of the state: Internet security and national security are diverging as well as converging. The second starts from the side of Internet governance: turning on their head usual analyses of legal globalization that start with the state, DeNardis describes laws and regulations of geopolitically recognized states as "bordered policy," with borders that are not always and in all ways contiguous with those of the networks being governed.

There are myriad types of states and theories about them (Held 1989), with social science interest in them rising and falling. As Jessop (2016, 1) notes, "here as in other fields, it seems that social scientists do not so much solve problems as get bored with them." The theories matter, though, because they not only reflect but also effect transformations in the nature of the state—so much so that Jessop insists conceptualizations of the state are among the core elements of the modern state, along with territory, state bureaucracy and resources, and the population. The same can be said for the ideas of governance and metagovernance. Jessop explores what he calls the historical semantics of the modern state, including the vocabulary used by theorists to describe and discuss the state and the roles of those conceptual frames in shaping the nature of the state and its practices. Using different language, Hofmann's chapter 12, on rhetorical functions that are served by decision-making structures and organizational forms, makes related arguments, as do ten Oever (2019) and Milan and ten Oever (2017) in their analyses of what activists and advocates do on the ground.

This section looks at two among the ways in which Internet governance is inextricably implicated in challenges to the state—treatment of borders, and the development of alternative governance forms. There is insufficient room here to explore additional important dimensions of these relationships and tensions as they pertain to rule of law and citizenship. Conceptual and methodological limits raised by such issues are discussed in the next section.

Borders of the State

Social and technical discussions of Internet governance have long been absorbed in the particular issues raised by borders, often but not only as they present in the form of jurisdictional dilemmas. On the social side, there has been work by legal scholars (classics include Burk 1997; Froomkin 1997; Johnson and Post 1996; and Zittrain 2005), many of whom felt the importance of the topic could not be overestimated; in Reidenberg's (2005) view, it would be on such issues that rule of law in Internet governance would rise or fall. The border-defining Internet governance issues that are the most familiar involve the domain name system (DNS) (Bradshaw and DeNardis 2018).

On the technical side, the first reference in the technical document series that is both medium for and documentation of the design process, the Requests for Comments (RFCs), to jurisdictional issues was when those responsible for design of what we now call the Internet were making the first international connection, to Norway, in 1972 (NORSAR n.d.). A decade later, it was argued that no technical restrictions on transborder email should be allowed that are any different in kind from those used with physical mail (the question involved encryption), and it was suggested that, when Internet design and architecture come into conflict with the law, governments should change their laws to solve the problem (RFC 828 [Owen 1982]). Other border issues that came up in the first decades, each launching ongoing design debates and efforts, included identity certification procedures (e.g., RFC 1114 [Kent and Linn 1989]), addressing (e.g., RFC 1218 [North American Directory Forum 1991]), and network security (e.g., RFC 1244 [Holbrook and Reynolds 1991]). By the close of the 1980s, discussion in RFCs included references to comparative legal scholarship (e.g., RFC 1135 [Reynolds 1989]).

The question of how to think about the boundaries of the geopolitical state in network terms has always been a challenge. Rutkowski (2011) provides an invaluable history of what we think of as cybersecurity principles, beginning with how they were first used in treaties dealing with postal systems and semaphore networks long before the digital era. Provisions included such things as exchanging network architecture and addressing information, data retention, authentication of messages of governments, and filtering of harmful messages. Among the principles to which almost every country in the world was a signatory at the time of his analysis are not only the proactive requirement to ensure communications security (a contemporary concern in the international legal community because of the question of how much a government needs to know regarding what flows through networks that cross its geographic territory), but also the rights to cut off all state connections with the international network and to cut off all private communications deemed dangerous "to the security of the state or contrary to its laws, to public order or to decency" (Rutkowski 2011, 15). Today we popularly refer to "rights to cut off" as the Internet kill switch (Vargas-Leon 2016).

Importantly, Rutkowski (2011) included a detailed description of what it took for him to collect all the pertinent documents from multiple institutional sources and to transfer the data from print and digital formats into a common digital format for analysis, processes that clearly took significant amounts of time and resources. By publicizing his data collection method and making the materials publicly available for use by others, Rutkowski urges us to do the same with other historical materials we need to study. In what other areas of Internet governance research would this kind of recuperation and sharing of primary research materials be of value? Should an institution take leadership in hosting such materials on behalf of the global community of Internet governance researchers?

International telecommunications network border issues have always been complex. Negotiations over international telecommunications networks in the late 19th and early 20th centuries revolved around such fundamental matters as the nature of corporations and what it means to be "foreign" (Zajácz 2019). For a long time, the US Federal Communications Commission operationally treated Canada and Mexico as domestic for regulatory purposes, and Alaska and Hawaii as foreign, because the agency found it more convenient to determine the boundaries by those of the specific network technology involved (land technologies to get to Canada and Mexico, underwater technologies to get to Alaska and Hawaii) than by geopolitics. Decisions about where to put the "border" between countries linked by a telecommunications circuit, necessary for the purposes of determining who paid whom for what when it comes to network flows in the world of telecommunications regulation as it had long been, were conceptual and negotiated (Frieden 1993), lending particular importance to national categorical decisions made regarding how to regulate the "converged" technological environment once digitization was at hand (Frieden 1984, 2004) from the international network perspective.

What is required to carry through on commitments made in the international treaties about electronic networks analyzed by Rutkowski (2011) is, of course, qualitatively and quantitatively more complex and thus difficult to understand and to successfully engage with in the contemporary environment. *Tallinn Manual 2.0* (Schmitt 2017) essentially suggests a new policy principle—the right of a state *not* to know what is going through networks that cross its territory or for which it has any other kind of responsibility, as a protection against needing to act on it in ways that could be self-destructive and/or politically unacceptable (Braman 2017). There are other areas in which developments provide limits to states, most notably with cybersecurity—something cannot be considered an action of national or homeland security concern unless it is an action against the state or the state, or an agent of the state, is involved. Ongoing questions for Internet governance researchers include asking how we can tell if something or someone is "of the state" in any given context or process or agent, and whether that is so by fiat, de jure or de facto. The problems are conceptual, empirical, and—perhaps, apparently, and/or potentially—contextual. How do ICANN processes affect perceptions of effective boundaries and expectations that governance must come from geopolitically recognized states in accord with rule of law? What are the effects of the DNS structure on regional, national, and local identities and community functions? What are the effective borders in cyberspace that ordinary users experience when engaging in activities of particular types, when they exhibit certain characteristics and/or when there are tensions between activities as bounded by geopolitical, network, or institutional (third-party intermediary) "jurisdictions"? How do users know when they are crossing each of these types of borders, and what matters to them when they do? To what extent are users aware when they cross borders on the Internet, and when does that matter to them?

Another path of Internet governance research involving borders became visible when the US government began surveilling information held on travelers' electronic devices at border crossings (for a period, irrespective of citizenship status) and related developments. At the beginning of the 21st century, a draft "PATRIOT Act 2" named the Domestic Security Enhancement Act was proposed that included reducing or eliminating some US citizenship rights for those who publicly expressed concern about the civil liberties and human rights dimensions of post-9/11 security and surveillance practices. It seems possible that a continuation of the Trump administration in the United States could result in dilution of citizenship rights on the basis of relationships with foreign nationals conducted through or facilitated by the Internet, providing a model that might well then be followed by other countries. What kinds of political and legal proposals are being put forward regarding treatment of information and communication that crosses borders, the relationship networks of which they are a part, and the identities of those who engage in cross-border flows and relationships in countries around the world, whether those are directed at the border gateway protocol within the network, censorship of certain types of content, or other control points?

Alternative Governance Forms

It was during the second of the stages in the history of the literature on changing states that we saw the rise of governance forms that are alternative to the state and in competition with it (Bevir and Rhodes 2011). The Internet and its governance have provided the affordances through which to develop multinational and transnational organizational forms of even greater scale, scope, flexibility, and power relative to states. The network makes it possible for such organizations to engage in "regulatory arbitrage,"

maximizing organizational ability to take advantage of resources and opportunities offered by geopolitically recognized entities (Froomkin 1999). The
work by Raymond and DeNardis (2015) on the varieties of multistakeholderism and Kulesza's (2018) analysis of its limits are foundational on this
subject for those studying Internet governance. Musiani (2016) points out
that the DNS itself can simultaneously be considered a technology and an
alternative governance institution.

It should not be surprising that as geopolitical governments privatize
formerly public functions and corporations simultaneously become increasingly governmental, the public and private sector entities as social technologies converge, yielding new organizational and decision-making forms.
Two literatures studying these developments have been particularly influential. Scholars in the law and society tradition study the ways in which
internal practices, programs, and rules serve governance functions (beyond
the more traditionally acknowledged and publicly aggressive approaches
of lobbying and revolving doors) (Mather 2013). In an example of such
work relevant to Internet governance research as defined here, Bamberger
and Mulligan (2015) studied corporate implementation of privacy law in
several countries. The second literature of importance is that on governmentality as theorized by Michel Foucault (1980, 1982, 1984). Cerny (2008)
provides one example of this type of analysis as applied to the looseness,
ambiguities, flexibility, uncertainties, and multiplicities of the Internet
governance environment, but Foucauldian ideas have influenced many,
including thinkers in mainstream political science (e.g., Jessop 2011). The
two approaches—law and society, and governmentality—are not in conflict
with each other, but whichever is used, these trends can make it more difficult for those studying Internet governance to pursue many types of questions because access to information becomes more difficult when it can
be claimed that data and its processing are proprietary, when the data are
lost altogether because what had been considered a government function is
abandoned, or when the data are unavailable for other reasons. Jørgensen
in chapter 13 provides important experience-based advice in this regard,
but here the most pressing challenges that Internet governance researchers
face are conceptual and methodological.

There are recurring aspirations to use the Internet to operate completely outside the reach of national and international law, whether that is
attempted physically, as in the Sealand experiment, or through some other

means. The current locus of such aspirations is the blockchain, though as Werbach (2018) informatively and persuasively argues, the blockchain will be most useful and is most likely to succeed if it is fully articulated within the law. That point is not limited to the blockchain, though how such efforts relate to the law will vary. In her analysis of internal attempts by Silk Road to prevent illegal online activity in pursuit of the anonymous networking entity's long-term success, Zajácz (2017) illustrates one reason this is so—the very protections for anonymity that allowed that network's illegal activities to continue as long as they did out of view of the state in turn undermined efforts by those who managed Silk Road to guarantee that participants in the marketplace would meet their obligations in the manner required for the trust needed to sustain the market. Across types of cybersecurity incidents, Wolff (2018) has found that it is much harder for those who engage in cyberattacks to use what they get once they bring it out of a database or network than it is for them to get in.

A seriously understudied type of Internet governance is that taking place within the autonomous systems the Internet, "a system of interconnected autonomous systems" (Tozal 2016), comprises. They are important control points (DeNardis 2012, 2014), can be categorized by the types of peering relationships they preferentially engage in (Tozal 2016), and can be mapped geographically as well as topologically, providing insight into additional ways in which such systems may be subject to political control (Yacobi-Keller et al. 2018). Some argue that the multiplicity of these systems justify describing the Internet as distributed rather than decentralized (Mathew 2016). When there are cyberattacks, they are leveled at autonomous systems, so Nur and Tozal (2018) suggest a method for measuring centrality that is based on how relatively critical a given autonomous system is to the network, or to society, as a whole. And that is about as far as the literature goes. Will we start thinking in terms of citizenship in autonomous systems, as the network unit of particular importance to user operations? Would going this route make it easier to think about governance tools that can be used in common irrespective of whether the autonomous system involved is public or private, or are different sets of governance tools still needed for the two types of subjects? With whom does thinking about autonomous systems from a governance perspective put ICANN in relation?

These trends come together with the increasing governmentalization of social media. Reversing the trend of what has been happening with citizenship, which has been becoming thinner and thinner, platforms are building from thin to thick: first currency is offered, then a means of engaging in transactions, then dispute resolution, and on. The announcement of Libra and its associated app that provides infrastructure for transactions and other financial matters from Facebook and collaborator corporations has gone the furthest in this regard by the time of writing. Geopolitical governance entities are insisting that the offerings undergo government review and approval before proceeding. What are the further concerns, if any, from the Internet governance perspective? To what extent, and how, are Internet governance decisions responsible should such offerings draw significant resources away from the state and toward the corporation that hosts the social media platform that will serve as the governance infrastructure, in turn contributing to the weakening of rule of law from the perspective of geopolitically recognized states?

Conceptual and Methodological Limits

Although, as discussed above, the general socio-technical point has by now been beautifully made, illustrated, and practiced, today's Internet governance research questions require attention to analytical dimensions not historically addressed. This in turn requires us to go further theoretically, conceptually and methodologically in many areas for which this this chapter does not have space, such as level and type of complexity and liability issues. Following, though, are some of the conceptual, theoretical, and methodological challenges.

Stages of Information Processing
We are most accustomed to thinking about speech issues in terms of single speech acts that involve identifiable and quantifiable (unitizable) expressions, flows of information, or sets of interactions. The sensemaking literature, and that which focuses on meaning making by an active audience, looks at how multiple flows come together at specific human conjunctures but still typically involve single steps. The problem of what it would take to effectively and meaningfully evaluate algorithmic decision-making, however, reminds us that in most, if not almost all, cases, it will actually be a stream of steps of information processing that yields the effect of any given one. How many steps back need to be known to be confident of the

provenance of informational or communicative content or of an action, whether by a human or daemon agent? Each information processing step can involve interactions with others that may themselves be complex systems involving diverse types of liability and levels of responsibility. Under what conditions is a specific agent to be held accountable for an act or communication resulting from interactions among many agents? We are beginning to hear calls for provenance of data regarding its information processing history from those seeking accountability and transparency of algorithms. How is such provenance most accurately and usefully determined? Who is to hold the records?

Conceptual and historical kernels of what it would take to achieve operationalizable answers to these questions for the purposes of current use within geopolitical legal systems and within the network exist. Treatment of the right to receive information as a necessary element of the right to communicate in the United States (and thus something that is also protected by the First Amendment) is one important example of an inherently multistep approach to thinking about the stages of information processing that need protection in order to achieve the constitutional goal. What new rights might be recognized as a result of further developing our ability to analyze algorithms for governance purposes? How might we come to articulate existing rights in new ways in order to map validly onto the digital environment in comprehensible ways? In what forms might such developments appear across the diverse institutions involved in Internet governance? What methods might be developed to identify when rights are being supported or being curtailed?

Analysis of information policy in US Supreme Court decisions of the 1980s (Braman 1988) found that essentially all involved some sort of reference to a model of an information production chain on which distinctions deemed to be of constitutional importance relied, whether implicitly or explicitly. A highly systematic effort to conceptualize information policy provisions across stages of the information production chain is underway by Rebecca MacKinnon's Ranking Digital Rights nonprofit organization (https://rankingdigitalrights.org). How can these and other extant models available for distinguishing among stages of information production chains be used in Internet governance research? Do we need to conceptualize additional stages of such chains, or conceptualize known stages differently, in order to adequately understand what is happening online for Internet governance purposes? What methodological tools can we develop to distinguish among stages?

Types of Information Processing

Reliance on information production chain models in the course of legal analysis means that distinctions among types of information processing are being made for governance purposes. When struggles over specific types of information processing ensue, often what is at stake is whether the kind of processing involved is speech or not, with the answer determining whether fundamental protections for human rights and civil liberties, such as those protecting free speech, apply. Legal debates over this as they specifically arise in the digital environment go back at least to the 1990s (Burk 2000). Battles over treating code as speech should be expected to continue (Coleman 2009; Petersen 2015). In the world of what is now called cyber operations, the question of how to distinguish cybercrime (matters of national law) from cyberwar (matters of international law) has proven so difficult that this new terminology was developed to refer to the two jointly for analytical purposes (Schmitt 2017).

Just how to draw distinctions among types of information processing in the online environment is at the heart of debates over governance of and by algorithms. Most of the literature in this terrain to date has started by identifying undesirable effects of the use of particular algorithms, such as bias or incitement. The work on information policy in constitutional law mentioned above distinguished between only two types of information processing—human and machinic, what we now refer to as algorithmic. Already in the 1980s it was clear both that that was far from adequate, and that going further would require significant intellectual effort. Now that the algorithmic moment is before us, it is time to invest in that effort. For each issue examined by an Internet governance researcher, what are the types of information processing involved, and what are the distinct governance problems attached to each? Is there a distinction between general-purpose and single-purpose information production chains that is useful for governance purposes? How do the policy tools used to address the same issue as it may arise across different stages of the information production chain vary? Under conditions in which it can be assumed that technological innovation will continue at a significant pace and highly competitively, is it possible to group types of processes together for evaluative and governance purposes with confidence that the range of types of information processing of concern, and their uses, are actually being taken into account? What methods can be developed to evaluate the efficacy of such policy tools?

Types of Speakers and Information Processors

Legal systems typically distinguish among types of communication speakers and receivers; examples include children versus adults, willing versus unwilling receiver, employee versus nonemployee, and spouse versus nonspouse. Analytical dimensions of long-standing interest include whether, and, if so, how, a speaker is anonymous; independent or autonomous (two related but distinct features); intelligent; and someone with whom there is a confidential relationship. In the digital environment, we have come additionally to think about whether the speaker is human or machinic, and we have come to think of speech as a distinct set of types of information processing. For those studying Internet governance in light of its importance vis-à-vis governance in general, it is particularly useful to approach the human versus machinic aspect through the lens of the question of who or what qualifies as a legal subject, and as a citizen.

Recognition of the distinction between human and machinic users—the latter referred to as daemons—was evident very early on in the history of Internet design (Braman 2011; McKelvey 2018). Each type of user required different things from the designers, some of whom communicated that they found the needs of daemons easier to understand than those of humans. It was acknowledged, however, that serving the needs of humans, odd as they might seem, led to solutions that were in many cases also significant improvements for the network. Artificial intelligence research focuses on how technologies can do things that humans do better than humans can do them. We are just beginning to see research on the separate question of how humans and daemons may differ when they engage in the same types of information processing. The story that deterrence games about nuclear war used computers rather than humans because it was found that humans would never launch nuclear warheads but computers would may be apocryphal, but it is telling even so. Vempaty et al. (2018) have found that humans and computers *do* differ when they engage in decision-making on at least some kinds of problems. This line of research is extremely important for those analyzing Internet governance processes. What kinds of differences in how the two types of information processors work need to be taken into account? How do these differences affect the values that are embedded in technology design or network architecture? How could such differences be presented in public and policy-making conversations in such a manner that they would make sense and could reasonably be the subject of decision-making?

The question of whether or not an entity is a legal subject arises both for individuals and for groups of people. Regarding the former, there is now a field of "robot law" (see, e.g., Calo, Froomkin, and Kerr 2016) that begins with the issue of personhood. When it comes to the latter, autonomous networks like WikiLeaks—not to be mistaken with the autonomous systems discussed above—are already proving problematic, as the US government found when it attempted to pursue that entity in the course of the Bradley (now Chelsea) Manning trial (Braman 2014). As social technologies continue to converge, producing new social forms, this problem is likely to arise repeatedly. The question of whether nonhumans can become legal subjects first came to the fore in the 1970s in the United States, when the issue was the environment. The history begins with a student, an origin story of a type familiar to Internet governance researchers: a law school student, Christopher Stone—son of the highly independent and influential investigative journalist I. F. Stone—sent his argument, since published as the book *Should Trees Have Standing?* (Stone 1974), to US Supreme Court Justice William O. Douglas, who in turn convinced the court to recognize that forests have standing before the law and thus have rights that could be legally protected. The suggestion that entities with artificial intelligence should be granted personhood was put forward before the Internet was commercialized and made available for general use (Solum 1991). Resonances between discussions of legal personhood for biological and machinic nonhuman entities continue (see, e.g., Solaiman 2017; Teubner 2006). What would change in Internet governance should daemons (which can include things like operating systems and network layers) and algorithms be granted legal personhood? How should the governance implications of such a fundamentally important development be taken into account when considering the question? Should all daemons and algorithms be treated alike from this perspective and, if not, what methods can be developed for drawing lines?

The literatures on citizenship dimensions of the Internet are large and diverse, as most broadly conceptualized beginning with political communication research on the impact of the Internet on political behaviors. John Perry Barlow (1996) famously issued a "Declaration of Independence" for those who see themselves first and foremost as citizens of cyberspace, and shortly afterward it became popular to think in terms of the network citizen (Hauben and Hauben 1997). A reading of the Internet RFCs from the perspective of what they say about legal, policy, and political matters over time shows the development of a sense of network citizenship that can conflict with geopolitical citizenship, creating problems for Internet governance (Braman 2013). This author's first exposure to the concept in the mass media was in a Singaporean newspaper in 2000, which used "network citizenship" as a term the readers were assumed to understand to refer to individuals who engaged in political activity online. Indeed, a quick search in summer of 2019 of the literature for "network citizenship" finds most of it is research on Asian countries. The development of China's social credit system (Liang et al. 2018) provides evidence of one direction in which network citizenship can evolve. MacKinnon (2012) argues for another, reviving the use of the concept in her call for development of a sense of global participation in decision-making for the Internet, done in a manner that effectively—not just rhetorically—protects human rights and civil liberties. How can tensions between geopolitical and network political citizenship be resolved? The history of the concept of citizenship is marked by the addition of dimensions considered to be inherent to the notion; do we need additional conceptual work to take into account the nature and implications of network citizenship? How does approaching Internet governance research through the lens of network citizenship affect the questions we ask and how we ask them? What does inclusion of both machinic and network citizenship require of the indicators we use to evaluate the nature and strength of democratic practices?

Governability and Irony

No one disputes that such matters as equity and fairness should be taken into account in technology design and network architecture. The challenge is that doing so may require more than we are now actually capable of in terms of fully understanding what is happening when highly complex systems multiply interact with each other across scales, geographies, uses, scopes, and contextual conditions. This is no apologia, nor a suggestion that demands for incorporating attention to such matters within design processes in as accountable a manner as possible shouldn't continue to be fought for—but it *is* an expression of humility in the face of what might actually be humanly possible when it comes to the governability of this environment, irrespective of by whom or for what purposes.

In the face of what so many political scientists now acknowledge to be the inevitability of governance failures, Jessop (2011) identifies three ways in which governance entities manage to not fail, to appear not to fail, or to, in essence, redefine what is meant by failure. An entity can achieve the appearance of successful governance by making sure that any problems appear beyond the spatiotemporal horizons (or perceptions of horizons) of a given set of players or forces. It can survive through a continual *fuite en avant*, continually escaping from one crisis by turning to another mode of policy making that will also fail. Or policy makers could engage in self-reflexive irony, even when they know they are likely to fail.

For Internet governance researchers, the first two of these are not within reach, but the third is. There are many ways of thinking about the job of an intellectual, but Clifford Geertz reminds us that, among other things, it is a way of life, and so are the methods we use:

> The various disciplines…are more than just intellectual coigns of vantage but are ways of being in the world, to invoke a Heideggerian formula, forms of life, to use a Wittgensteinian, or varieties of noetic experience, to adapt a Jamesian. In the same way that Papuans or Amazonians inhabit the world they imagine, so do high energy physicists or historians of the Mediterranean in the age of Phillip II…. To set out to deconstruct Yeats's imagery, absorb oneself in black holes, or measure the effect of schooling on economic achievement is not just to take up a technical task but to take on a cultural frame that defines a great part of one's life. (Geertz 1982, 24)

The researcher *is* the methodological instrument irrespective of any other tools used (Lindlof and Taylor 2018). When we are talking about method, we are talking about who we are.

The irony of Internet governance research is that, as most broadly defined, Internet governance itself may not exist in any of a number of senses. Policies we do put in place may not be effective. What makes sense technically may not work politically, socially, or legally, and vice versa. It is becoming difficult to separate the Internet as the subject of governance of human communications from a wide variety of other types of processes and from the material environment itself. The subject (and subjects) may be too complex to be governable at all, or it may not even exist as a legal subject. Political processes we thought were stable enough to support continued incremental social, economic, political, and cultural change are turning out not to be. Put together, all of these make Internet governance research a problem in metagovernance.

Whether or not Internet governance exists, we have a right to it—or may choose to go on as if we believe we do. Researchers are involved with lots of "ologies," from epistemology and methodology on. Deontology is among them.

Note

1. Sandra Braman, from *The One Verse City* (Eugene, OR: Wolf Run Books, 1974).

References

Abbate, J. (2017). What and where is the Internet? (Re)defining Internet histories. *Internet Histories, 1*(1–2), 8–14.

Antonelli, C. (1988). The emergence of the networked firm. In Cristiano Antonelli (Ed.), *New information technology and industrial change: The Italian case* (pp. 13–32). Dordrecht, Netherlands: Springer.

Antonelli, C. (2017). Digital knowledge generation and the appropriability trade-off. *Telecommunications Policy, 41*, 991–1002.

Archer, M. S. (1982). Morphogenesis versus structuration: On combining structure and action. *The British Journal of Sociology, 33*(4), 455–483.

Archer, M. S. (1988). Towards theoretical unification: Structure, culture and morphogenesis. *Culture and agency: The place of culture in social theory* (pp. 247–307). Cambridge, UK: Cambridge University Press.

Bache, I., Bartle, I., Flinders, M., & Marsden, G. (2015). *Multilevel governance and climate change: Insights from transport policy.* Pickering & Chatto.

Bambauer, J. R. (2017). Dr. Robot. *UC Davis Law Review, 51*, 383–398.

Bamberger, K. A., & Mulligan, D. K. (2015). *Privacy on the ground: Driving corporate behavior in the United States and Europe.* Cambridge, MA: MIT Press.

Barlow, J. P. (1996). A declaration of the independence of cyberspace. *Duke Law & Technology Review, 18*(1), 5–7.

Beckett, S. (1953/2012). *The unnameable.* New York, NY: Grove Press.

Bevir, M. (2011). Governance as theory, practice, and dilemma. In M. Bevir (Ed.), *The Sage handbook of governance* (pp. 1–16). Thousand Oaks, CA: Sage.

Bevir, M., & Rhodes, R. A. W. (2011). The stateless state. In M. Bevir (Ed.), *The Sage handbook of governance* (pp. 203–217). London, UK: Sage Publications.

Bevir, M., & Rhodes, R. A. W. (2016). The "3Rs" in rethinking governance: Ruling, rationalities, and resistance. In M. Bevir & R. A. W Rhodes (Eds.), *Rethinking governance: Ruling, rationalities and resistance* (pp. 1–21). New York, NY: Routledge.

Bovaird, T. (2005). Public governance: Balancing stakeholder power in a network society. *International Review of Administrative Sciences, 71*(2), 217–228.

Bowker, G. (1994). *Science on the run: Information management and industrial geophysics at Schlumberger, 1920–1940.* Cambridge, MA: MIT Press.

Bradshaw, S., & DeNardis, L. (2018). The politicization of the Internet's domain name system: Implications for Internet security, universality, and freedom. *New Media & Society, 20*(1), 332–350.

Braman, S. (1974). *The one verse city.* Eugene, OR: Wolf Run Books.

Braman, S. (1988). *Information policy and the United States Supreme Court* (Unpublished dissertation). University of Minnesota. Dissertation #8823521; Proquest document ID 303670545.

Braman, S. (1995). Policy for the net and the Internet. *Annual Review of Information Science and Technology (ARIST), 30,* 5–75.

Braman, S. (2002). Informational meta-technologies and international relations: The case of biotechnologies. In J. Rosenau & J. P. Singh (Eds.), *Information technologies and global politics: The changing scope of power and governance* (pp. 91–112). Albany: State University of New York Press.

Braman, S. (2004a). The processes of emergence. In *The emergent global information policy regime* (pp. 1–11). Houndsmills, UK: Palgrave Macmillan.

Braman, S. (2004b). The emergent global information policy regime. In *The emergent global information policy regime* (pp. 12–37). Houndsmills, UK: Palgrave Macmillan.

Braman, S. (2004c). Where has media policy gone? Defining the field in the twenty-first century. *Communication Law and Policy, 9*(2), 153–158.

Braman, S. (2005). The micro- and macroeconomics of information. *Annual Review of Information Science and Technology (ARIST), 40,* 3–52.

Braman, S. (2006). *Change of state: Information, policy, and power.* Cambridge, MA: MIT Press.

Braman, S. (2011). The framing years: Policy fundamentals in the Internet design process, 1969–1979. *The Information Society, 27*(5), 295–310.

Braman, S. (2013). The geopolitical vs. the network political: Internet designers and governance. *International Journal of Media & Cultural Politics, 9*(3), 277–296.

Braman, S. (2014). "We are Bradley Manning": Information policy, the legal subject, and the WikiLeaks complex. *International Journal of Communication, 8*, 2603–2618.

Braman, S. (2015). The state of cloud computing policy. In C. Yoo & J.-F. Blanchette (Eds.), *Regulating the cloud: Policy for computing infrastructure* (pp. 279–288). Cambridge, MA: MIT Press.

Braman, S. (2017). The medium as power: Information and its flows as acts of war. In C. George (Ed.), *Communicating with power* (pp. 3–22). Bern, Switzerland: Peter Lang, International Communication Association Theme Book Series.

Browne, M. (1997). The field of information policy, I: Fundamental concepts. *Journal of Information Science, 23*(4), 261–275.

Bruce, R. R., Cunard, J. P., & Director, M. D. (1986). *From telecommunications to electronic services: A global spectrum of definitions, boundary lines, and structures.* Boston, MA: Butterworth.

Brügger, N., Goggin, G., Milligan, I., & Schafer, V. (2017). Introduction: Internet histories. *Internet Histories, 11*(1–2), 1–7.

Burk, D. (1997). Jurisdiction in a world without borders. *Virginia Journal of Law and Technology, 1*(3), 1522–1687.

Burk, D. (2000). Patenting speech. *Texas Law Review, 79*, 99–162.

Calo, R., Froomkin, A. M., & Kerr, I. (Eds.). (2016). *Robot law.* Edward Elgar.

Cerny, P. G. (2008). The governmentalization of world politics. In E. Kofman & G. Youngs (Eds.), *Globalization: Theory and practice* (3rd ed., pp. 221–237). London, UK: Continuum.

Clark, D. D. (2018). *Designing an Internet.* Cambridge, MA: MIT Press.

Cohen, J. (2012). *Configuring the networked self: Law, code, and the play of everyday practice.* New Haven, CT: Yale University Press.

Coleman, G. (2009). Code is speech: Legal tinkering, expertise, and protest among free and open source software developers. *Cultural Anthropology, 24*(3), 420–454.

Coletta, C., & Kitchin, R. (2017, July–December). Algorhythmic governance: Regulating the "heartbeat" of a city using the Internet of Things. *Big Data & Society*, 1–17.

Cowhey, P. F. (1990). The international telecommunications regime: The political roots of regimes for high technology. *International Organization, 44*(2), 169–199.

Crawford, S. P. (2007). The Internet and the project of communication law. *UCLA Law Review, 55*, 359–407.

DeLanda, M. (1991). *War in the age of intelligent machines*. New York, NY: Zone Books.

DeNardis, L. (2012). Hidden levers of Internet control. *Information, Communication & Society, 15*(5), 720–738.

DeNardis, L. (2014). *The global war on Internet governance*. New Haven, CT: Yale University Press.

DeNardis, L., & Musiani, F. (2016). Governance by infrastructure. In F. Musiani, D. L. Cogburn, L. DeNardis, & N. S. Levinson (Eds.), *The turn to infrastructure in Internet governance* (pp. 1–23). Houndsmills, UK: Palgrave Macmillan.

Dezalay, Y., & Garth, B. G. (1996). *Dealing in virtue: International commercial arbitration and the construction of a transnational legal order*. Chicago, IL: University of Chicago Press.

Drezner, D. W. (2009). The power and peril of international regime complexity. *Perspectives on Politics, 7*(1), 65–70.

Edwards, L., & Veale, M. (2017). Slave to the algorithm? Why a "right to an explanation" is probably not the remedy you are looking for. *Duke Law & Technology Review, 16*(1), 18–84.

Epstein, D., Katzenbach, C., & Musiani, F. (2016). Doing Internet governance: Practices, controversies, infrastructures, and institutions. *Internet Policy Review, 5*(3). doi:10.14763/2016.3.435

Foucault, M. (1980). *Power/knowledge* (C. Gordon, Ed.). New York: Pantheon Books.

Foucault, M. (1982). The subject and power. *Critical Inquiry, 8*(4), 777–795.

Foucault, M. (1984). Space, knowledge and power. In P. Rabinow (Ed.), *The Foucault reader* (pp. 239–256). New York, NY: Pantheon Books.

Frieden, R. M. (1984). The international application of the Second Computer Inquiry. *Michigan Yearbook of International Legal Studies, 5*, 189–218.

Frieden, R. (1993). International toll revenue division: Tackling the inequalities and inefficiencies. *Telecommunications Policy, 17*(3), 221–233.

Frieden, R. M. (2004). The FCC's name game: How shifting regulatory classifications affect competition. *Berkeley Technology Law Journal, 19*(4), 1275–1314.

Froomkin, M. A. (1997). The Internet as a source of regulatory arbitrage. In B. Kahin & C. Nesson (Eds.), *Borders in cyberspace* (pp. 129–163). Cambridge, MA: MIT Press.

Froomkin, M. A. (1999). Of governments and governance. *Berkeley Technology Law Journal, 14*, 617–633.

Geertz, C. (1982). The way we think now: Toward an ethnography of modern thought. *Bulletin of the American Academy of Arts and Sciences, 35*(5), 24.

Giddens, A. (1984). *The constitution of society: Outline of the theory of structuration.* Berkeley: University of California Press.

Gurumurthy, A., & Chami, N. (2016). Internet governance as "ideology in practice"— India's "Free Basics" controversy. *Internet Policy Review, 5*(3). doi:10.14763.2016.3.431

Hauben, M., & Hauben, R. (1997). *Netizens: On the history and impact of Usenet and the Internet.* Los Alamitos, CA: IEEE Computer Society Press.

Held, D. (1989). *Political theory and the modern state.* Cambridge, UK: Polity Press.

Holbrook, J., & Reynolds, J. (Eds.). (1991). *Site security handbook.* RFC 1244. Retrieved from the Internet Engineering Task Force website: https://tools.ietf.org/html/rfc1244

Horwitz, R. B. (1989). *The irony of regulatory reform: The deregulation of American telecommunications.* Oxford, UK: Oxford University Press.

Jessop, B. (2004). Multi-level governance and multi-level metagovernance. In I. Bache & M. Finders (Eds.), *Multi-level governance* (pp. 49–74). Oxford, UK: Oxford University Press.

Jessop, B. (2011). Metagovernance. In M. Bevir (Ed.), *The Sage handbook of governance* (pp. 106–123). Thousand Oaks, CA: Sage.

Jessop, B. (2016). *The state: Past, present, and future.* Cambridge, MA: Polity Press.

Johnson, D., & Post, D. (1996). Law and borders: The rise of law in cyberspace. *Stanford Law Review, 48*(5), 1367–1402.

Kaye, D. (2019). *Speech police: The global struggle to govern the Internet.* New York, NY: Columbia Global Reports.

Kent, S., & Linn, J. (1989). *Privacy enhancement for Internet electronic mail: Part II— Certificate-based key management.* RFC 1114. Retrieved from the Internet Engineering Task Force website: https://tools.ietf.org/html/rfc1114

Krasner, S. D. (Ed.). (1983). *International regimes.* Ithaca, NY: Cornell University Press.

Kulesza, J. (2018). Balancing privacy and security in a multistakeholder environment: ICANN, WHOIS, and GDPR. *The Visio Journal, 3*, 48–58.

Law, J. (1992). Notes on the theory of the actor-network: Ordering, strategy, and heterogeneity. *Systems Practice, 5*(4), 379–393.

Liang, F., Das, V., Kostyuk, N., & Hussain, M. M. (2018). Constructing a data-driven society: China's social credit system as a state surveillance infrastructure. *Policy & Internet, 10*(4), 415–453.

Libra Association Members. (n.d.). *An introduction to Libra*. Libra White Paper. Retrieved from https://libra.org/en-US/white-paper/

Lindlof, T. R., & Taylor, B. (2018). *Qualitative communication research methods* (4th ed.). Thousand Oaks, CA: Sage Publications.

MacKinnon, R. (2012). *Consent of the networked: The worldwide struggle for Internet freedom*. New York, NY: Basic Books.

Martine, B. J. (1992). *Indeterminacy and intelligibility*. Albany: State University of New York Press.

Massumi, B. (2007). Potential politics and the primacy of preemption. *Theory & Event, 10*(2). doi:10.1353/tae.2007.0066

Mather, L. (2013). Law and society. In R. E. Goodin (Ed.), *The Oxford handbook of political science*. Oxford, UK: Oxford University Press. doi:10.1093/oxfordhb/97801 99604456.013.0015

Mathew, A. J. (2016). The myth of the decentralised Internet. *Internet Policy Review, 5*(3). doi:10.14763/2016.3.425

McKelvey, F. (2018). *Internet daemons: Digital communications possessed*. Minneapolis: University of Minnesota Press.

McLuhan, E., & McLuhan, M. (1992). *Laws of media: The new science*. Toronto, Canada: University of Toronto Press.

Merrill, K. (2016). Domains of control: Governance of and by the domain name system. In F. Musiani, D. L. Cogburn, L. DeNardis, & N. S. Levinson (Eds.), *The turn to infrastructure in Internet governance* (pp. 89–106). New York, NY: Palgrave Macmillan.

Meuleman, L. (2008). *Public management and the metagovernance of hierarchies, networks, and markets: The feasibility of designing and managing governance style combinations*. The Hague, Netherlands: Physica-Verlag.

Milan, S., & ten Oever, N. (2017). Coding and encoding rights in Internet infrastructure. *Internet Policy Review, 6*(1). doi:1.14763/2017.1442

The Moscow Times. (2019, February 12). Russia moves to grant government the power to shut down the Internet, explained. *The Moscow Times*. Retrieved from https://www.themoscowtimes.com/2019/02/12/russia-moves-grant-government -power-shut-down-internet-explained-a64470

Mueller, M. (2004). ICANN and INTELSAT: Global communication technologies and their incorporation into international regimes. In Sandra Braman (Ed.), *The emergent global information policy regime* (pp. 62–85). Houndsmills, UK: Palgrave Macmillan.

Musiani, F. (2015). Practice, plurality, performativity, and plumbing: Internet governance research meets science and technology studies. *Science, Technology, & Human Values, 40*(2), 272–286.

Musiani, F. (2016). Alternative technologies as alternative institutions: The case of the domain name system. In F. Musiani, D. L. Cogburn, L. DeNardis, & N. S. Levinson (Eds.), *The turn to infrastructure in Internet governance* (pp. 73–87). New York, NY: Palgrave Macmillan.

NORSAR. (n.d.). NORSAR and the Internet. Retrieved May 22, 2010, from http://www.norsar.no/pc-5-30-NORSAR-and-the-Internet.aspx

North American Directory Forum, The. (1991). *A naming scheme for c=US*. RFC 1218. Retrieved from the Internet Engineering Task Force website: https://tools.ietf.org/html/rfc1218

Nur, A. Y., & Tozal, M. E. (2018). Identifying critical autonomous systems in the Internet. *The Journal of Supercomputing, 74*(10), 4965–4985.

Nye, J. S. (2014). *The regime complex for managing global cyber activities*. Global Commission on Internet Governance Paper Series (Paper no. 1). Centre for International Governance Innovation/Chatham House. Retrieved from https://www.cigionline.org/sites/default/files/gcig_paper_no1.pdf

OECD. (2018). *OECD science, technology and innovation outlook 2018: Adapting to technological and societal disruption*. Paris, France: Author.

Owen, K. (1982). *Data communications: IFIP's international "network" of experts*. RFC 828. Retrieved from the Internet Engineering Task Force website: https://tools.ietf.org/html/rfc828

Petersen, J. (2015). Is code speech? Law and the expressivity of machinic language. *New Media & Society, 17*(3), 415–431.

Pool, I. de Sola. (1983). *Technologies of freedom*. Cambridge, MA: Belknap Press.

Rajnish, P., Ranjan, B. S., Mridula, P., Shrutilka, B., Neha, A., & Divya, S. (2017). Role of Internet governance in achievement of Sustainable Development Goals. *International Journal of Research in Social Sciences, 7*(9), 574–577.

Raymond, M., & DeNardis, L. (2015). Multistakeholderism: Anatomy of an inchoate global institution. *International Theory, 7*(3), 571–616.

Reidenberg, J. (2005). Technology and Internet jurisdiction. *University of Pennsylvania Law Review, 153*, 1951–1974.

Reynolds, J. (1989). *The helminthiasis of the Internet*. RFC 1135. Retrieved from the Internet Engineering Task Force website: https://tools.ietf.org/html/rfc1135

Russell, A. L. (2017). Hagiography, revisionism & blasphemy in Internet histories. *Internet Histories, 1*(1–2), 15–25.

Rutkowski, A. (2011). Public international law of the international telecommunication instruments: Cyber security treaty provisions since 1850. *Info, 13*(1), 13–31.

Schmitt, M. N. (Ed.). (2013). *Tallinn manual.* Cambridge, UK: Cambridge University Press.

Schmitt, M. N. (Ed.). (2017). *Tallinn manual 2.0.* Cambridge, UK: Cambridge University Press.

Solaiman, S. M. (2017). Legal personality of robots, corporations, idols and chimpanzees: A quest for legitimacy. *Artificial Intelligence and Law, 25*(2), 155–179.

Solum, L. B. (1991). Legal personhood for artificial intelligences. *North Carolina Law Review, 70,* 1231.

Star, S. L., & Ruhleder, K. (1996). Steps toward an ecology of infrastructure: Design and access for large information spaces. *Information Systems Research, 7*(1), 111–134.

Stone, C. D. (1974). *Should trees have standing?* Palo Alto, CA: William Kaufmann.

Suskind, R. (2004, October 17). Faith, certainty and the presidency of George W. Bush. *The New York Times Magazine.* Retrieved from www.nytimes.com/2004/10/17/magazine/17BUSH.html

Tass. (2019, May 1). Putin signs law on reliable Russian Internet. *Tass.* Retrieved from http://tass.com/politics/1056750

ten Oever, N. (2019). Productive contestation, civil society, and global governance: Human rights as a boundary object in ICANN. *Policy & Internet, 11*(1), 37–60.

Teubner, G. (2006). Rights of non-humans? Electronic agents and animals as new actors in politics and law. *Journal of Law and Society, 33*(4), 497–521.

Theall, D. F. (1999). *James Joyce's techno-poetics.* Toronto, Ontario: University of Toronto Press.

Tozal, M. E. (2016). The Internet: A system of interconnected autonomous systems. In *2016 Annual IEEE Systems Conference (SysCon)* (pp. 1–8). IEEE.

Tribe, L. H. (1985). Constitutional calculus: Equal justice or economic efficiency? *Harvard Law Review, 98*(3), 592–621.

Vargas-Leon, P. (2016). Tracking Internet shutdown practices: Democracies and hybrid regimes. In F. Musiani, D. L. Cogburn, L. DeNardis, & N. S. Levinson (Eds.), *The turn to infrastructure in Internet governance* (pp. 167–188). Houndsmills, UK: Palgrave Macmillan.

Vempaty, A., Varshney, L. R., Koop, G. J., Criss, A. H., & Varshney, P. K. (2018). Experiments and models for decision fusion by humans in inference networks. *IEEE Transactions on Signal Processing, 66*(11), 2960–2971.

Werbach, K. (2018). *The blockchain and the new architecture of trust.* Cambridge, MA: MIT Press.

Wolff, J. (2018). *You'll see this message when it is too late: The legal and economic aftermath of cybersecurity breaches.* Cambridge, MA: MIT Press.

Yacobi-Keller, U., Savin, E., Fabian, B., & Ermakova, T. (2018). Towards geographical analysis of the autonomous system network. *International Journal of Networking and Virtual Organizations, 21*(3), 379–397.

Zajácz, R. (2017). Silk Road: The market beyond the reach of the state, *The Information Society, 33*(1), 1–12.

Zajácz, R. (2019). *Reluctant power: Networks, corporations, and the struggle for global governance in the early 20th century.* Cambridge, MA: MIT Press.

Zittrain, J. L. (2005). *Jurisdiction.* St. Paul, MN: Foundation Press.

3   Inventing Internet Governance: The Historical Trajectory of the Phenomenon and the Field

Milton L. Mueller and Farzaneh Badiei

This chapter tracks the emergence of "Internet governance" as a label, a field of research and academic study, and a real-world arena where stakeholders and interest groups clash and cooperate. We try to look at all three of them simultaneously—label, field of study, and set of practices and institutions—focusing on the interplay among them over time. We have chosen to begin our assessment of the field of Internet governance on the basis of when the term started to be consciously recognized as a phenomenon and labeled as such.

Some may argue that some form of Internet governance was occurring before this; they might, for example, begin with the US Department of Defense's ARPANET and would characterize hammering out some of the early design principles of internetworking as Internet governance. An even broader approach, developed by Sandra Braman in chapter 2, tries to situate Internet governance in the convergence of computing and communication technologies in the 1950s and 1960s and the globalization of communication networks in the 1970s and 1980s. And while it is true that the Internet entered a policy context shaped by these processes, it is also true that the policy and governance of integrated services digital network (ISDN), or cross-border data flows over private leased telecommunication circuits, cannot be characterized as Internet governance. The Internet had its own distinctive protocols that posed unique governance problems. The Internet also evolved its own standards development organizations and governance institutions, such as the Internet Engineering Task Force (IETF) and address registries, that were outside the established institutions of global telecommunications governance.

Historical periodizations are neither correct nor incorrect; they are more or less suited to specific purposes. Our purpose is not to track communications policy in general but to reveal the trajectory of Internet governance as a distinct arena of public policy and global governance. By doing that, we hope to assess the usefulness and longevity of the term "Internet governance" as the label for this field. It makes sense, therefore, to start with the time period after the Internet protocols had been developed and implemented and their use by the public had reached the point at which it posed problems that had to be resolved in public arenas through legal, political, and institutional means. This did not happen until the Internet was open to widespread public use and was recognized as something that could be or should be subject to public governance.

Our chapter contributes to the overarching theme of this book by analyzing the evolution of the field of Internet governance studies. We show how Internet governance research and scholarly participation appeared at stages of the Internet's development, the emergence of distinct Internet governance problems, and how and why they became important research topics. This can help scholars identify emerging topics that relate to Internet governance. We have also identified how various disciplines got involved with Internet governance research. For example, early on, the Internet governance field was mostly rooted in legal studies, but as the Internet became more widespread, its governance started to intersect with other fields such as political science and international relations. We briefly mention the emergence of theories of Internet governance, new research methods, and the use of new research tools to analyze Internet-generated data.

Phase One (1993–1997): Discovery and Exceptionalism

The term "Internet governance," as far as we can tell, does not appear in any scholarly or news articles before 1995 (Kowack 1995). But the words "governance" and "law," on the one hand, and "cyberspace," "Internet," and "the Net," on the other, started to be used in close association with each other several years before that, roughly from 1993 on (see, for example, Braman 1995). That period corresponds to the emergence of the Internet as a mass public medium. That emergence was contingent on three events: the development and adoption of the World Wide Web protocol from 1989–1993, the publication of freely downloadable web browser software after 1991, and the privatization of the Internet backbone and its opening to commercial use by the US National Science Foundation in 1995. Together, these developments made the Internet accessible to ordinary businesses and end users. (For an early newspaper account of this transition, see Lewis [1994].)

The most notable and interesting feature of the earliest Internet governance discourse is its vigorous debate on conceiving of cyberspace as its own place, or what some have called Internet exceptionalism. John Perry Barlow (1996) is usually put forward as the paragon of exceptionalism, but typically this is done to discredit the idea on the cheap. Barlow is a convenient strawman for antiexceptionalists because he was a rock lyricist, not a social science scholar, and his *Declaration* was a manifesto and rallying cry, not a carefully developed, theoretically grounded argument.

But contemporary to Barlow, legal scholars were having a far more serious debate about the extent to which cyberspace was a terrain that should develop its own rules and institutions. David Johnson and David Post were also exceptionalists (Johnson and Post 1996), and they were among the first to ask explicitly, "How shall the net be governed?" (Johnson and Post 1997). Focusing on the Internet's operational reliance on voluntary transborder cooperation, they developed an argument for decentralized, emergent law as an alternative to traditional hierarchical, state-centric control. The word "governance," which was also gaining credence in UN documents at the same time (UNDP 1997), was a softer term than "government" and implied a more polycentric order.

Johnson and Post's argument is often unfairly equated with a cruder, technological determinist argument that the Internet is inherently resistant to state control. But the emergent law argument was normative, not positive: it asserted that the Internet could and probably *should* follow a new model of nonnational governance, not that it necessarily *would* be governed in that way.

Even before Johnson and Post, in what is perhaps the earliest law review article on cyberspace governance, Hardy (1994) also opted for the exceptionalist camp. Like others in this period, he does not use the term "Internet governance" but refers to cyberspace law, or the customs and rules associated with its governance. His paper asks whether we should treat (the whole of) cyberspace as a separate jurisdiction. While inspired by law and economics scholars and judges, Hardy also refers to multistakeholder collaboration and bottom-up rulemaking. He rejects established intergovernmental organizations as unsuitable for resolving transjurisdictional disputes on the Internet and instead puts forward the idea of having a world cyber court. He calls for "international cooperation among a wide array of groups" for the court, which seems to imply a multistakeholder governance system. Similarly, Tim Wu addressed the issue of cyberspace sovereignty—not the sovereignty of states, but of cyberspace itself—in 1997 (Wu 1997). Wu's treatment contained a remarkably prescient discussion of the application of international relations theory to cyberspace governance. As with Hardy, cyberspace governance, not Internet governance, is the dominant label. It was, as the next section explains, the formation of the Internet Corporation for Assigned Names and Numbers (ICANN) that cemented the term "Internet governance" into place.

Aside from the broad discourse about the new world created by cyberspace, legal scholars were attuned to the less sweeping but still novel and intellectually exciting problems raised by the commercialization of the Internet. Novel legal issues were posed by the nascent online economy:

- Trademark law and domain names (Burk 1995)
- Intermediary responsibility and copyright protection (Boyle 1996; Hardy 1994; Samuelson 1996)
- Censorship and filtering of Internet content (Resnick 1997; Resnick and Miller 1996)
- Private contract law versus public law and jurisdictional conflict (Perritt 1996)

As the fascination with cyberlaw studies burgeoned, conservative legal scholars reacted against it, insisting that there was no new cyber jurisdiction or territory or that calls for a new jurisdiction constituted a call for "cyber-anarchy" (Goldsmith 1998). Cyberlaw was ridiculed as tantamount to the study of "the law of the horse" (Easterbrook 1996). Such an effort, a prominent legal scholar proclaimed, was bound to be "shallow and to miss unifying principles" (Easterbrook 1996, 207).

Easterbrook was claiming—wrongly as it turned out—that technology did not or could not transform societal interactions sufficiently to justify a specific analysis of how law related to the technology. This evolved into a debate on the sociology of technology. Legal scholars were asking whether Internet technology altered laws and institutions or were incorporated into established legal principles, just as science and technology studies had for some years before been defined by an ongoing debate between realists who emphasize the autonomous or deterministic effect of technologies and social constructivists who focus on how societal factors shape the final form taken by technical systems. Lessig's (1999a) code-as-law argument was, in essence, an attempt by a legal scholar to reinvent the wheel of social studies of technology.

A key feature of this period is that the Internet governance field is rooted in legal studies. The literature is already addressing issues in international relations, institutional theory, science and technology studies, global governance, and the role of global civil society, but legal scholars are doing most of the work (Kowack [1995] is a rare exception). In retrospect, almost every aspect of current issues in Internet governance, including the nexus between national security and cybersecurity (Arquilla and Ronfeldt 1993), had been posed in some form in these initial years. Yet in this first phase, North America is overwhelmingly the center of research and writing. Very little attention is paid to the implications of Internet growth and policy to other nations or cultures or to how Internet governance might play into interstate rivalries.

Phase Two (1996–2003): ICANN *Über Alles*

The term "Internet governance" came to prominence in 1996–1999, when it became associated with the struggle to create a new institution to take over global coordination of Internet domain names and IP addresses. The formation of ICANN took center stage in Internet governance discourse and research, and the term "Internet governance" became widely used to describe this area.[1] Ironically, during the period in which "Internet governance" becomes widely used as the label for the conflicts over control of the root of the domain name system (DNS) and the policy battles associated with domain-name trademark conflicts, the term is also actively resisted by many involved in the names and numbers debates, especially the technical community stakeholders.[2] In what turns out to be a losing battle, they insist that "governance" is a misleading term because the Internet Assigned Numbers Authority (IANA) is not the Internet and governance of domain name and IP address resources is merely technical management, not governance or policy. A seminal book from this period, which published the results of a 1996 conference of academics and practitioners around the problem of institutionalizing the IANA, was deliberately titled *Coordinating the Internet* and did not refer to governing it (Kahin and Keller 1997). And yet, almost all results of searching the LexisNexis Academic database for the term "Internet governance" in this period relate to the controversies over control of IANA and the formation of ICANN.

Research and writing during this period shifted away from more abstract exceptionalism debates to the question of building a real governance institution. But exceptionalism was often an unstated assumption, in that few actors in a position of influence wanted the Internet to be subsumed under existing intergovernmental regimes. Who, then, would control a global, centralized institutional framework to coordinate domain name and address assignment? How would this institution be structured? How would it be made accountable? Fulfilling the expectations and norms of the exceptionalists, the encounter with those problems culminated in an institutional innovation, the ICANN (Mueller 2002).

The formation of the ICANN regime resolved conflicts over property rights that had been created by attempts to appropriate new global technical resources (primarily domain names). It also addressed the coordination problems posed by managing critical Internet resources in a manner that would retain global compatibility. As Wolfgang Kleinwächter noted, ICANN was a "silent subversive" because of the way it altered the role of states in global governance (Kleinwächter 2001). Indeed, for a time ICANN was hailed as a paradigm of new forms of governance ushered in by the networked age (Ahlert 2001; Hofmann 2005; Levinson 2002). Others, however, while recognizing its novelty, mounted strong challenges to the model's legality and legitimacy (Froomkin 2000; Weinberg 2000).

ICANN was controversial because it was a private nonprofit corporation unilaterally delegated by the United States to be the global authority over the root of the domain name and Internet address spaces and empowered to resolve key public policy problems through the issuance of private contracts. These contracts were a means of addressing competition policy issues in the commercial market for domain names, domain-name trademark conflicts, the allocation of Internet addresses, and related problems. Klein and several others explore another interesting aspect of the ICANN experiment— namely, its early attempt to use global, democratic elections to keep its board of directors accountable (Klein 2001; Palfrey 2004; Weinberg 2001).

Given the overarching *problematique* of the relationship between internetworking and national sovereignty, the role of governments in ICANN's formation has always been a topic of interest, and we begin to see political scientists drawn into Internet governance research. Volker Leib and Daniel Drezner, for example, examined EU–US interactions in the initial negotiations over ICANN (Drezner 2007; Farrell 2003; Leib 2002). In the real world of institutions, this *problematique* led to the creation and gradual empowerment of ICANN's Governmental Advisory Committee (GAC). The GAC was a strange beast: simultaneously a mini-intergovernmental organization composed of representatives of nation-states and an organ of a private California nonprofit public benefit corporation offering nonbinding advice to the organization (Weinberg 2011).

Phase Three (2003–2009): World Summit on the Information Society and Internet Governance Forum

In the 2003–2009 phase, Internet governance becomes fully recognized as a domain of global governance, and the boundaries of what is considered Internet governance expand beyond ICANN. In the business world, this phase also marks the rise of social media platforms around global corporations such as Facebook, Twitter, and Google. Large-scale global intermediaries residing on the Internet transform the context in which traditional communication policy issues are debated. But the key turning point in both research field formation and the actual practice of governance was the World Summit on the Information Society (WSIS).

During the WSIS, which lasted from 2002 to 2005, ICANN became the provocation for international clashes over the US role in Internet governance and the position of state and nonstate actors in shaping it. Using Google search counts as a metric, we find that the period of the WSIS corresponds to the high point of "Internet governance" as a search term with currency among the web-using public (figure 3.1). Awareness of Internet governance was raised in the developing world and especially among diplomats and government ministries, who had to learn what Internet governance (and perhaps even what the Internet) was. It led to the creation, in 2004, of the UN Working Group on Internet Governance (WGIG), which was charged with developing a working definition of the term. In the early stages of the WSIS process, definitional debates centered on the distinction between a narrow approach, encompassing only ICANN-related functions, and a broad definition, seeming to include anything and everything related

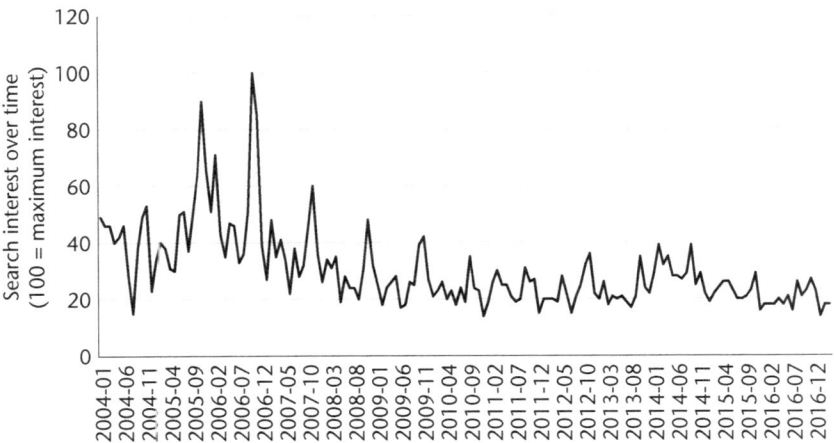

Figure 3.1
Occurrence of "Internet governance" as a search term worldwide by year.

to information and communication technologies. Both extremes missed the mark. Confining concepts of Internet governance to ICANN was justified by the undeniable fact that DNS and IP governance was central to the actual functioning of the global Internet, but it was evident that other processes also governed the global Internet, such as standardization bodies, trade in services agreements, World Intellectual Property Organization treaties, law enforcement activities related to cybercrime, and so on. On the other hand, any attempt to stretch "Internet governance" to include matters like the construction of physical telecommunications infrastructure, spectrum management, open standards, national e-government initiatives, and the like was simply based on an uncritical attempt to conflate all forms of information and communication technology with the Internet.

In its agreed definition, the WGIG expanded the meaning of Internet governance beyond ICANN, applying the term to any and all "shared principles, norms, rules, decision-making procedures, and programmes that shape the evolution and use of the Internet" (Drake 2005; MacLean 2004). The definition obviously drew on Krasner's (1983) canonical definition of international regimes but, reflecting the enlarged role of nonstate actors in managing the Internet, noted that these shared processes involve not just governments but business and civil society as well.

The WGIG-WSIS definition ratified the position of nonstate actors in Internet governance and put many of the traditional problems of communication and information policy within its frame. Reinforcing this trend, in their confrontation with states, ICANN and its defenders found it useful to emphasize the open and multistakeholder nature of ICANN processes. What had been described as private sector leadership or self-regulation in the early days of ICANN's formation was now repackaged as the multistakeholder model. Intergovernmental organizations such as the United Nations and intergovernmental processes such as the WSIS had to be opened up to civil society and the private sector. Whereas, before, ICANN's most powerful actors had scorned or marginalized civil society stakeholders, they now embraced them as evidence of their relative openness and the superiority of its private-sector-led governance model. A new line of research opened around multistakeholder governance, and Internet governance researchers began to look at other domains such as the environment for precedents and at the preexisting literature on transnational governance networks (Cave et al. 2007; DeNardis and Raymond 2013; Levinson and Smith 2008; Sørenson and Torfing 2007).

The literature on the WSIS is large and of uneven quality but contains many important insights into international institutions, the participation of civil society in global governance, the role of the United States, and of course Internet governance itself. One of the ironies of the WSIS is that it was supposed to address the full range of communication-information policy but ended up becoming almost entirely focused on ICANN and Internet governance. A good descriptive analysis of the WSIS process from the standpoint of a traditional civil society development advocate and UN system insider can be found in Souter (2007). Hans Klein (2004) provides a valuable analysis of the politics of the WSIS placed in the context of UN summits. Wolfgang Kleinwächter (2004) and Marc Raboy (2004) provide additional participant accounts of the WSIS process, while Mueller (2010) examines the post-WSIS Internet governance landscape and emphasizes the continuing tension between networks and states as forms of governance.

WSIS created a new set of expectations regarding the ability of civil society actors to participate in global Internet governance processes (Padovani and Tuzzi 2004; O'Siochru 2004; Raboy 2004). While it often disappointed stronger advocates of participatory democracy and failed to resolve the debates about ICANN and the US unilateral role in Internet governance, WSIS did create a new institutional vehicle for carrying on discussion and debate around those issues: the multistakeholder Internet Governance Forum (IGF). There is a huge amount of policy literature and occasional papers around the IGF but very little deep scholarly analysis. Malcolm (2008) carefully traces the developments of the IGF's first two years and offers a normative analysis of how it can be reformed to fulfill the promise of multistakeholder governance.

Another immediate result of WSIS was the formation of an academic network specifically devoted to Internet governance research: the Global Internet Governance Academic Network (GigaNet) in 2006. In early 2006, during the formative stages of the new UN IGF, emails and conversations among a core group of academics led to a conclusion that within the post-WSIS environment there was no natural home for Internet governance research and education. They decided to create their own independent academic platform, introducing a sometimes awkward separation from the civil society nongovernmental organizations with whom they had been connected during the WSIS process. GigaNet has held annual symposia showcasing Internet governance research concurrently with the IGF every year since the first IGF in 2006.

The dialogue and writing about the appropriate definition of Internet governance was not entirely settled by WSIS but continues to this day. To some, Internet governance includes only those technical, legal, regulatory and policy problems that arise as a direct consequence of the involved parties' mutual use of the Internet protocols to communicate. Laura DeNardis provides a clearly reasoned basis for distinguishing between studies of Internet content and usage and the governance of the Internet per se. "Issues of Internet governance relate to Internet-unique technical architecture rather than the larger sphere of information and communication technology design and policy" (DeNardis, 2014, 19). Another issue is that governance in cyberspace is so distributed and indirect that it is often unclear where authority to govern lies. Van Eeten and Mueller (2013) initiated a critical scholarly debate about whether IGF and other forums in which Internet governance is discussed can be considered Internet governance at all. That debate is taken up by Hofmann, Katzenbach, and Gollack (2016), who attempt to introduce a distinction between governance and regulation.

Whether the IGF constitutes governance or a mere talk shop, it seems to have set in motion a process of institutional isomorphism, with multiple national and regional Internet governance forums following in its wake (Epstein and Nonnecke 2016). Studying these forums, Epstein and Non-necke develop a distinction between substantive and performative multistakeholder governance of the Internet. Although the IGF is more of a performative multistakeholder Internet governance process, they argue that the regional and national Internet governance forums may become the link between the UN IGF and local Internet rulemaking.

While WSIS captured the attention of scholars who self-identified as Internet governance researchers and analysts, a whole new set of governance problems was brewing in the largely noninstitutionalized space formed by transnational Internet services and commerce. There was growing awareness of the power of states to shape Internet governance, led by an oft-cited work by Goldsmith and Wu (2006) arguing that states are still in control and everything will return to normal. At the same time, other scholars emphasized the ways the Internet had altered the nature of global governance of information and communication and that new forms of governance were developing.

One key area of development was Internet content regulation—that is, blocking and filtering. As nation-states gradually learned how to filter and block websites from outside their jurisdiction as an attempt to maintain sovereign control of information—and users and external actors explored ways to circumvent those barriers—an important empirical body of literature grew up around efforts by scholars to track and understand these practices. Ronald Deibert's Citizen Lab at the University of Toronto developed technical tools and methods for the systematic global analysis of Internet filtering by states (Deibert et al. 2008, 2010, 2012). Computer scientists also joined the fun, using automated Internet measurement techniques to collect data on censorship and circumvention practices (Leberknight et al. 2010; Wolfgarten 2005). Growing attention was also paid to the way private intermediaries regulated content (Wagner 2016).

A school of research focused on the economics of information security also developed during this period. While this began with a more traditional information security focus (Anderson 2001), the growth of cyberspace and the movement of ever-more social capabilities onto the Internet meant that these researchers increasingly intersected with cybersecurity and Internet governance issues, although they rarely used those labels to describe their work (Anderson and Moore 2007; Moore 2008). The new field was based on the insight that the Internet's security problems are not simply technical but are driven by the economic incentives of actors and firms. Work in this area feeds into policy discourse by analyzing, for example, the cost-benefit trade-offs of Internet service providers' efforts to secure their networks and customers, the assignment of liability to software producers or Internet service providers, the impact of network externalities or tragedies of the commons, and the ways in which markets interact with government action in response to security problems.

Intermediary liability had been considered an Internet governance issue since the earliest days of the commercial Internet (Hardy 1994), but discussions of the topic became more prevalent with increases in Internet usage and the growing profitability of Internet intermediaries. The rise of online market intermediaries (e.g., eBay and Amazon), search engines such as Google, and social networks such as Facebook and Twitter brought to the fore new policy issues around topics such as defamation, copyright and counterfeit goods, and e-commerce (see, e.g., Mann and Belzley 2005). In this period, the immunity of intermediaries from liability was extensively discussed (Lemley 2007). This discussion continued into the fourth phase of the field's evolution, especially with efforts to pressure social media platforms to identify and take down terrorist content and the European Court of Justice decision on the right to be forgotten, which forced intermediaries to delink materials from their search engine results at the request of a claimant.

The tension between the Internet's capacity to quickly and easily share digital content, on the one hand, and the protection of copyright and trademarks, on the other, provided one of the major flashpoints of policy conflict and research. Media giants and national and international rights protection collectives had the resources to pursue a copyright and trademark protectionist strategy on a global basis. Incumbent copyright holders became some of the strongest advocates for imposing policing and enforcement responsibilities on Internet service providers and social media platforms (Bridy 2010; Horten 2012; Mueller, Kuehn, and Santoso 2012). The DNS also became an arena where rights holders sought preemptive protections for rights to names (Froomkin 2002; Galloway and Komaitis 2005).

If the rise of the Internet has mobilized copyright and trademark owners, it also sparked a new social movement that pushed in the opposite direction (Benkler 2006; Boyle 1997; Lessig 2005). This access to knowledge (A2K) movement was inspired by institutional analyses that view a commons as a desirable governance model (Kranich 2004). Open, nonproprietary access to information was seen as especially appropriate because consumption of information is nonrivalrous. The movement had its origins in the developers of free and open source software, who pioneered new contractual mechanisms deliberately designed to prevent informational resources from being privately appropriated (O'Mahony 2003; Raymond 2001; Stallman 2002; Weber 2004). By the middle of the first decade of the 2000s, this actor network had become a full-fledged social movement that melded the free software movement with critics of the patent system in drugs and biotechnology and opponents of copyright and trademark maximalism. The A2K movement, like its opponent, was transnational in scope and self-consciously took its cause into international organizations (notably the World Intellectual Property Organization) and national legislatures (May 2007). These two forces had a historic collision in 2011–2012 over two proposed laws in the United States that would have implemented domain-name blocking mechanisms similar to China's in the service of copyright enforcement (Benkler et al. 2015; Sell 2013).

Phase Four (2010–): Surveillance, Securitization, and Alignment

In the fourth phase, ongoing as we write, issues of surveillance, privacy, and cybersecurity have become increasingly central to Internet governance politics and research. The Internet is going through a process of securitization (Cavelty 2007; Deibert and Rohozinski 2010), which further reinforces the linkages between the nation-state and Internet governance. As this happens, interest in and explicit mentions of "Internet governance" in the fields of political science and international relations grow exponentially (figure 3.2). In the research explicitly focused on the national and transnational power implications of the Internet's vulnerabilities, the term "security" no longer refers to more narrow technical forms of security but starts to mean exactly what it does in mainstream international relations research. Internet governance research now overlaps with studies of war and interstate conflict, deterrence, foreign policy, espionage, terrorist groups, and the threat to critical infrastructures that might be posed through cyberspace vulnerabilities. These concerns are often explicitly linked to the international diplomatic and policy conflicts over Internet governance (e.g., Segal 2016). Well-known international relations (IR) scholars such as Joseph Nye, who were unfamiliar

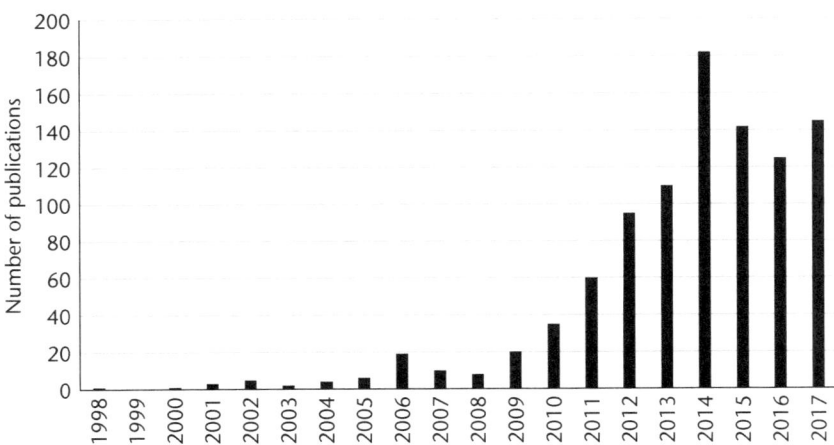

Figure 3.2
Occurrence of "Internet governance" in journals in ProQuest political science database by year.

with the decade of Internet governance research preceding the involvement of political scientists, reinvented certain themes, characterizing Internet governance as a "regime complex" (Nye 2014). There is even a systematic application to cyberspace of the classic IR concepts of the security dilemma (Buchanan 2016) and deterrence (Gartzke, Lindsay, and Nacht 2014).

The Edward Snowden revelations, which exposed internal documents about the pervasive global surveillance of the US National Security Agency, were a watershed in this process. The classified documents confirmed the surprisingly large scale and scope of digital surveillance, rekindling long-standing debates about privacy, encryption, and the powers of the state relative to the individual. But they also solidified the link between military and national security and the Internet; reinforced notions that one government, the United States, was preeminent or hegemonic in cyberspace; opened the veil on how the United States leveraged the extensive data collection that its private intermediaries gather about their users (Lyon 2014); and undermined US moral authority in Internet rights and norms (Farrell and Finnemore 2013). Snowden generated new forms of "data nationalism" from allies and rivals alike (Chander and Lê 2015).

In addition to sparking a vibrant debate about the possibility of a Balkanized or fragmented Internet (Drake, Cerf, and Kleinwächter 2016; Hill 2012;

Mueller 2017), the Snowden revelations contributed to a major institutional change in Internet governance—namely, the IANA transition (Becker 2019; Mueller 2014; Scholte 2016). US control of ICANN via the IANA functions contract had been controversial since the early days of WSIS. The post-Snowden crisis of legitimacy finally prompted the United States to relinquish its control of the DNS root and its contractual control of the IANA functions, owing to fears that many countries would defect from the ICANN-led Internet governance regime. The transition also brought in its wake major reforms in ICANN's accountability arrangements (Kruger 2015).

Despite its drift toward state-centric approaches, the cybersecurity literature does overlap with Internet-governance-related research on networked and multistakeholder governance. Empirical research on the actual mechanisms of cybersecurity production reveals a great deal of private action and collective action by Internet service providers, standard-setting organizations, and governments and law enforcement agencies (Asghari et al. 2015). Traditional hierarchical state action is the exception rather than the rule (Kuerbis and Badiei 2017; Schmidt 2014).

Surveillance and privacy were also key factors in the civilian debate over policy responses to the ubiquity of social media. While traditional privacy advocates adjusted their norms and policy ideas to the new conditions (Trottier 2016), neo-Marxists spoke of surveillance capitalism as a new type of economy (Zuboff 2015, 2019). The governance of data protection loomed ever larger in Internet governance, as the right to be forgotten, the breakdown of the US-Europe safe harbor agreement, and Europe's General Data Protection Regulation transformed the regulatory environment for major platforms offering free services in exchange for the value of users' data (Bennett and Raab 2017).

Data-localization laws were also analyzed as a digital trade issue. Legal scholars studied how data localization can affect digital trade and cross-border data flow, framing it as a trade barrier without achieving the desired privacy and security on the Internet (Chander and Lê 2015). Others took a different approach, discussing the legitimacy of data-localization laws by countries such as Brazil, India, China, and Russia. Assessing the tension between data localization and free trade, Selby argues that data-localization laws engage both Internet governance and international trade law (Selby 2017).

Another recent development in the field of Internet governance and trade is tech-nationalism. The US attack on Chinese telecommunications equipment manufacturer Huawei pushes the next generation of mobile tele-communications into the fray of nation-state rivalry. Russia has also taken an approach to the Internet that is explicitly nationalist and sovereign (Stadnik 2019). These actions against free trade are taken in the name of national security and cybersecurity. As nations ramp up more tech-nationalistic tendencies, the effect of sanctions and trade protectionism on technologies that are used to operate the Internet might become a more prevalent research topic.

Methodologically, the field trends toward a growing use of computer science–based measurement techniques. Some of the most promising new research comes from scholars who exploit the information-generating tools of the Internet itself to compile and analyze data about the Internet. One sees reverse engineering of spyware by Deibert's Citizen Lab and efforts by researchers to gain control of the command-and-control infrastructure of a botnet. Scholars of the economics of security such as Tyler Moore and Michel van Eeten also are increasingly able to mine huge computer-generated databases of phishing activities, routing information, spam sources, and the like. It is possible to imagine a broader diffusion and further evolution of these methods to bear more directly on the problems of Internet governance. This implies a synthesis of technical knowledge and social science that is still too rare.

Conclusion

Internet governance as a label and field of study has undergone a remarkable evolution over the last 20 years. Once a term applied narrowly to debates around the control of the DNS root—and hotly contested and rejected as a label even then—it has now become an accepted designator for a broad range of policy issues, institutional developments, and geopolitical phenomena. The need to define the term even generated a special UN working group as part of a UN summit process.

Figure 3.3 provides an overview of how often "Internet governance" occurs in academic and research publication databases. The data are shown for four different disciplines: law (from LexisNexis Academic), economics (from EconLit database), political science (from ProQuest Political Science database, which includes international relations journals), and sociology (from ProQuest Sociological Abstracts). From this we can see that the field of law has shown the most consistent and sustained interest in Internet

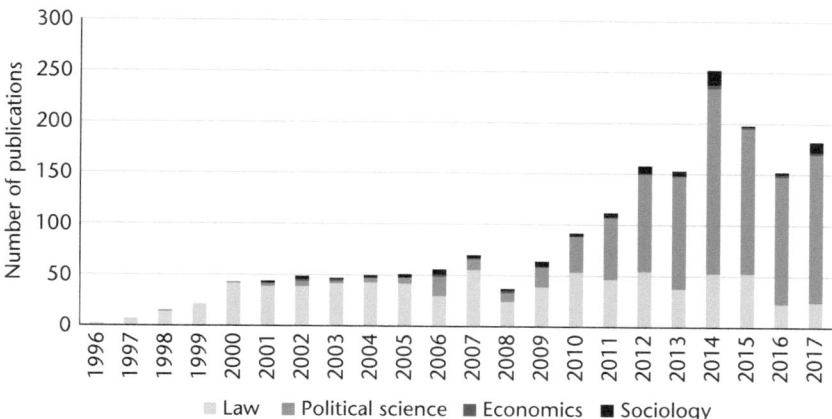

Figure 3.3
Occurrence of "Internet governance" in academic journals by year.

governance as a topic, ramping up quickly from 2 occurrences in 1996 to 49 in 2000 and maintaining a rate of about 40–50 publications per year since then. Since 2010, however, political science and international relations publications have dominated the field in terms of the sheer number of publications. We attribute this to the rise of cybersecurity as a policy and research preoccupation and the growing perception of Internet governance as a geopolitical issue. The amount of interest generated by cybersecurity and cyberspace governance in Internet research and political science fields testifies to the depth of the challenge Internet connectivity and computer technology poses to the traditional form of the state.

Economics and sociology, by way of contrast, have shown the least interest in the topic. Economics articles on the topic typically trickle in at 2 or 3 a year. Sociology averages around 4 articles a year, although it, like political science, shows more activity since 2010, notably a peak of 14 articles in 2014 and another spike of 10 in 2017. This is not, of course, because economists and sociologists have no interest in the social transformations caused by the Internet—it is simply that few of them frame their concerns as Internet governance. They are more focused on the evolution of Internet industries and the users and uses of the new media, respectively.

We conclude by raising an interesting but possibly uncomfortable question about the future of the field. In an article, Michel van Eeten claims that the rise of the Internet of Things and pervasive computing signal[s] the disappearance of the distinction between devices with and without connectivity and computing capabilities. Without that distinction, it also becomes less meaningful to think about cybersecurity governance as a space with a certain structural coherence. (2017, 437)

Should the governance of Internet-connected medical devices, for example, be considered Internet governance or part of health policy? Are autonomous vehicles handled as Internet governance or transportation policy? As the Internet and connected devices become ubiquitous, it is possible that most of the governance questions will be confronted and resolved in sector-specific ways that fall outside the realm of Internet governance. If this happens, ironically, it may be that the definition and scope of Internet governance once again reverts to the narrow realm of the Internet's naming, addressing, and routing infrastructure.

Notes

1. Lessig (1999b, 1407) used the term "governance" in a very general sense, and when explaining the word he mentioned the procedures for domain name registration as an Internet governance issue.

2. The NTIA green paper that eventually led to ICANN clearly reflects this tension in the following passage: "This discussion draft, shaped by the public input described above, provides notice and seeks public comment on a proposal to improve the technical management of Internet names and addresses. It does not propose a monolithic structure for Internet governance. We doubt that the Internet should be governed by one plan or one body or even by a series of plans and bodies." "Improvement of Technical Management of Internet Names and Addresses; Proposed Rule," February 20, 1998, docket number 980212036-8036-01, retrieved from https://www.ntia.doc.gov/federal-register-notice /1998/improvement-technical-management-internet-names-and-addresses-proposed-.

References

Ahlert, C. (2001). Democr@tic-Global-Governance.net: ICANN als Paradigma neuer Formen Internationaler Politik [Democr@tic-global-governance.net: ICANN as a paradigm of new forms of international government in cyberspace]. *Internationale Politik und Gesellschaft, 1*, 66–78.

Anderson, R. (2001). Why information security is hard—an economic perspective. *Proceedings of the 17th Annual Computer Security Applications Conference.* Retrieved from http://www.cl.cam.ac.uk/~rja14/Papers/econ.pdf

Anderson, R., & Moore, T. (2007, January). *Information security economics—and beyond.* Paper presented at the Fourth Bi-Annual Conference on the Economics of the Software and Internet Industries, Toulouse, France. Retrieved from http://www.cl.cam.ac.uk/~rja14/Papers/econ_crypto.pdf

Arquilla, J., & Ronfeldt, D. (1993). Cyberwar is coming! *Comparative Strategy, 12*(2), 141–165.

Asghari, H., van Eeten, M. J., & Bauer, J. M. (2015). Economics of fighting botnets: Lessons from a decade of mitigation. *IEEE Security & Privacy, 13*(5), 16–23.

Barlow, J. P. (1996). *A declaration of the independence of cyberspace.* Retrieved from https://www.eff.org/cyberspace-independence

Becker, M. (2019). When public principals give up control over private agents: The new independence of ICANN in Internet governance. *Regulation and Governance, 13*(4), 561–576.

Benkler, Y. (2006). *The wealth of networks: How social production transforms markets and freedom.* New Haven, CT: Yale University Press.

Benkler, Y., Roberts, H., Faris, R., Solow-Niederman, A., & Etling, B. (2015). Social mobilization and the networked public sphere: Mapping the SOPA-PIPA debate. *Political Communication, 32*(4), 594–624.

Bennett, C., & Raab, C. D. (2017). Revisiting "the governance of privacy." *SSRN.* Retrieved from https://papers.ssrn.com/sol3/papers.cfm?abstract_id=2972086

Boyle, J. (1996). *Shamans, software and spleens: Law and the construction of the information society.* Cambridge, MA: Harvard University Press.

Boyle, J. (1997). A politics of intellectual property: Environmentalism for the net? *Duke Law Journal, 47*, 87.

Braman, S. (1995). Policy for the net and the Internet. *Annual Review of Information Science and Technology, 30*, 5–75.

Bridy, A. (2010). Graduated response and the turn to private ordering in online copyright enforcement. *Oregon Law Review, 89*, 81.

Buchanan, B. (2016). *The cybersecurity dilemma: Hacking, trust and fear between nations.* Oxford, UK: Oxford University Press.

Burk, D. (1995). Trademarks along the infobahn: A first look at the emerging law of cybermarks. *Richmond Journal of Law and Technology, 1*(1), 1–6. Retrieved from https://scholarship.richmond.edu/cgi/viewcontent.cgi?referer=https://scholar.google.com/&httpsredir=1&article=1003&context=jolt

Cave, J., Marsden, C., Klautzer, L., Levitt, R., van Oranje, C., Rabinovich, L., & Robinson, N. (2007, March 5). Responsibility in the global information society: Towards multi-stakeholder governance. *SSRN.* Retrieved from https://ssrn.com/abstract=2142027

Cavelty, M. D. (2007). *Cyber-security and threat politics: US efforts to secure the information age*. London, UK: Routledge.

Chander, A., & Lê, U. (2015). Data nationalism. *Emory Law Journal, 64*, 677.

Deibert, R., Palfrey, J. G., Rohozinski, R., & Zittrain, J. (Eds.). (2008). *Access denied: The practice and policy of global Internet filtering*. Cambridge, MA: MIT Press.

Deibert, R., Palfrey, J. G., Rohozinski, R., & Zittrain, J. (Eds.). (2010). *Access controlled: The shaping of power, rights and rule in cyberspace*. Cambridge, MA: MIT Press.

Deibert, R., Palfrey, J. G., Rohozinski, R., & Zittrain, J. (Eds.). (2012). *Access contested: Security, identity and resistance in Asian cyberspace*. Cambridge, MA: MIT Press.

Deibert, R. J., & Rohozinski, R. (2010). Risking security: Policies and paradoxes of cyberspace security. *International Political Sociology, 4*(1), 15–32.

DeNardis, L. (2014). *The global war for Internet governance*. New Haven, CT: Yale University Press.

DeNardis, L., & Raymond, M. (2013). Thinking clearly about multistakeholder Internet governance. *GigaNet: Global Internet Governance Academic Network, Annual Symposium 2013*. Retrieved from https://papers.ssrn.com/sol3/papers.cfm?abstract_id=2354377

Drake, W. J. (Ed.). (2005). *Reforming Internet governance: Perspectives from the working group on Internet governance*. New York, NY: UN ICT Task Force.

Drake, W. J., Cerf, V., & Kleinwächter, W. (2016, January). *Internet fragmentation: An overview*. Future of the Internet Initiative White Paper. Retrieved from World Economic Forum website: http://www3.weforum.org/docs/WEF_FII_Internet_Fragmentation_An_Overview_2016.pdf

Drezner, D. (2007). *All politics are global: Explaining international regulatory regimes*. Princeton, NJ: Princeton University Press.

Easterbrook, F. H. (1996). Cyberspace and the law of the horse. *University of Chicago Legal Forum, 1996*, 207–216.

Epstein, D., & Nonnecke, B. M. (2016). Multistakeholderism in praxis: The case of the regional and national Internet Governance Forum (IGF) initiatives. *Policy & Internet, 8*, 148–173. doi:10.1002/poi3.116

Farrell, H. (2003). Constructing the international foundations of e-commerce—the EU-US safe harbor arrangement. *International Organization, 57*(2), 277–306.

Farrell, H., & Finnemore, M. (2013). The end of hypocrisy: American foreign policy in the age of leaks. *Foreign Affairs, 92*(6), 22–26.

Froomkin, A. M. (2000). Wrong turn in cyberspace: Using ICANN to route around the APA and the Constitution. *Duke Law Journal, 50*(1), 17–186.

Froomkin, A. M. (2002). ICANN's UDRP: Its causes and partial cures. *Brooklyn Law Review, 67*(3), 605–718.

Galloway, J., & Komaitis, K. (2005). Like Alice in Wonderland: Applying EC competition principles in the case of domain names. *Journal of Information, Law and Technology, 2005*(2 & 3). Retrieved from https://warwick.ac.uk/fac/soc/law/elj/jilt/2005_2 -3/galloway-komaitis

Gartzke, E., Lindsay, J., & Nacht, M. (2014, March). *Cross-domain deterrence: Strategy in an era of complexity.* Paper presented at the International Studies Association Annual Meeting, Toronto, Canada.

Goldsmith, J. (1998). Against cyber-anarchy. *University of Chicago Law Review, 65*(4), 1199–1250.

Goldsmith, J., & Wu, T. (2006). *Who controls the Internet? Illusions of a borderless world.* New York, NY: Oxford University Press.

Hardy, I. T. (1994). The proper legal regime for "cyberspace." *University of Pittsburgh Law Review, 55*, 993–1055.

Hill, J. F. (2012). Internet fragmentation: Highlighting the major technical, governance and diplomatic challenges for US policy makers. *SSRN.* Retrieved from https:// papers.ssrn.com/sol3/papers.cfm?abstract_id=2439486

Hofmann, J. (2005). Internet Governance: Zwischen staatlicher Autoritat und privater Koordination. *Internationale Politik und Gesellschaft, 3*, 10–29.

Hofmann, J., Katzenbach, C., & Gollack, K. (2016). Between coordination and regulation: Finding the governance in Internet governance. *New Media and Society, 19* (9), 1406–1423.

Horten, M. (2012). *The copyright enforcement enigma: Internet politics and the 'telecoms package.'* UK: Palgrave MacMillan.

Johnson, D. R., & Post, D. (1996). Law and borders: The rise of law in cyberspace. *Stanford Law Review, 48*, 1367.

Johnson, D. R., & Post, D. (1997). And how shall the net be governed? A meditation on the relative virtues of decentralized, emergent law. In B. Kahin & J. H. Keller (Eds.), *Coordinating the Internet* (pp. 62–91). Cambridge, MA: MIT Press.

Kahin, B., & Keller, J. H. (Eds.). (1997). *Coordinating the Internet.* Cambridge, MA: MIT Press.

Klein, H. (2001). Global democracy and the ICANN elections. *Info, 3*(4), 255–257.

Klein, H. (2004). Understanding WSIS: An institutional analysis of the UN world summit on the information society. *Information Technology and International Development, 1*(3–4), 3–14.

Kleinwächter, W. (2000). ICANN between technical mandate and political challenges. *Telecommunications Policy, 24*(6), 553–565.

Kleinwächter, W. (2001). The silent subversive: ICANN and the new global governance. *Info, 3*(4), 259–278.

Kleinwächter, W. (2004). Beyond ICANN vs ITU? How WSIS tries to enter the new territory of Internet governance. *Gazette, 66*(3–4), 233–251.

Kowack, G. (1995). Internet governance and the emergence of global civil society. *IEEE Communications Magazine, 35*(5), 52–57.

Kranich, N. (2004). *The information commons: A public policy report* (No. 4). New York, NY: Free Expression Policy Project, Brennan Center for Justice, NYU School of Law.

Krasner, S. D. (1983). *International regimes*. Ithaca, NY: Cornell University Press.

Kruger, L. G. (2016, September). The future of Internet governance: Should the United States relinquish its authority over ICANN? (CRS Report R44022). *Congressional Research Service*. Retrieved from https://fas.org/sgp/crs/misc/R44022.pdf

Kuerbis, B., and Badiei, F. (2017). Mapping the cybersecurity institutional landscape. *Digital Policy, Regulation and Governance, 19*(6), 466–492.

Leberknight, C., Chiang, M., Poor, H. V., Wong, F. (2010, December 31). *A taxonomy of Internet censorship and anti-censorship*. Draft Version. Retrieved from http://www.princeton.edu/~chiangm/anticensorship.pdf

Leib, V. (2002). ICANN-EU can't: Internet governance and Europe's role in the formation of the Internet Corporation for Assigned Names and Numbers (ICANN). *Communication Abstracts, 25*(6), 755–909.

Lemley, M. A. (2007). Rationalizing Internet safe harbors. *Journal of Telecommunications and High Technology Law, 6*, 101.

Lessig, L. (1999a). *Code and other laws of cyberspace*. New York, NY: Basic Books.

Lessig, L. (1999b). Open code and open societies: Values of Internet governance. *Chicago-Kent Law Review, 74*, 1405.

Lessig, L. (2005). *Free culture: The nature and future of creativity*. New York, NY: Penguin.

Lewis, P. H. (1994, October 24). U.S. begins privatizing Internet's operations. *The New York Times*.

Levinson, N. S. (2002). Internet governance and institutional change. *The Tocqueville Review/La Revue Tocqueville, 23*(2), 125–141.

Levinson, N., & Smith, H. (2008, August 28). *The Internet governance ecosystem: Assessing multistakeholderism and change*. Paper presented at the American Political Science Association 2008 Annual Meeting, Hynes Convention Center, Boston. Retrieved

from http://195.130.87.21:8080/dspace/bitstream/123456789/1020/1/The%20inter
net%20governance%20ecosystem%20assessing%20multistakeholderism%20and%20
change.pdf

Lyon, D. (2014). Surveillance, Snowden, and big data: Capacities, consequences, critique. *Big Data & Society, 1*(2). doi:10.1177/2053951714541861

MacLean, D. (Ed.). (2004). *Internet governance: A grand collaboration.* New York, NY: UN ICT Task Force.

Malcolm, J. (2008). *Multi-stakeholder governance and the Internet Governance Forum.* Wembley, Australia: Terminus Press.

Mann, R. J., and Belzley, S. R. (2005). The promise of Internet intermediary liability. *William and Mary Law Review, 47,* 239.

May, C. (2007). The World Intellectual Property Organization and the development agenda. *Global Governance: A Review of Multilateralism and International Organizations, 13*(2), 161–170.

Moore, T. (2008). *Cooperative attack and defense in distributed networks* (Report No. UCAM-CL-TR-718). Cambridge, UK: University of Cambridge Computer Laboratory.

Mueller, M. (2002). *Ruling the root: Internet governance and the taming of cyberspace.* Cambridge, MA: MIT Press.

Mueller, M. (2010). *Networks and states: The global politics of Internet governance.* Cambridge, MA: MIT Press.

Mueller, M. (2014). Detaching Internet governance from the state: Globalizing the IANA. *Georgetown Journal of International Affairs,* 35–44.

Mueller, M. (2017). *Will the Internet fragment? Sovereignty, globalization and cyberspace.* London, UK: Polity.

Mueller, M., Kuehn, A., & Santoso, S. M. (2012). Policing the network: Using DPI for copyright enforcement. *Surveillance & Society, 9*(4), 348–364.

Nye, J. S. (2014). *The regime complex for managing global cyber activities.* Global Commission on Internet Governance Paper Series (Paper no. 1). Centre for International Governance Innovation/Chatham House. Retrieved from https://www.cigionline.org/sites/default/files/gcig_paper_no1.pdf

O'Mahony, S. (2003). Guarding the commons: How community managed software projects protect their work. *Research Policy, 32*(7), 1179–1198.

O'Siochru, S. (2004). Civil society participation in the WSIS process: Promises and reality. *Continuum: Journal of Media & Cultural Studies, 18*(3), 330–344.

Padovani, C., & Tuzzi, A. (2004). Global civil society and the world summit on the information society: Reflections on global governance, participation and the changing scope of political action. *Social Science Research Council*. Retrieved from https://ecpr.eu/Filestore/PaperProposal/268396ec-ddf8-4df3-bd32-763d314d6c65 .pdf

Palfrey, J. (2004). The end of the experiment: How ICANN's foray into global Internet democracy failed. *Harvard Journal of Law & Technology, 17*(2)

Perritt, H. H., Jr. (1996). Jurisdiction in cyberspace. *Villanova Law Review, 41*, 1.

Raboy, M. (2004). The WSIS as a political space in global media governance. *Continuum: Journal of Media & Cultural Studies, 18*(3), 345–359.

Raymond, E. S. (2001). *The cathedral and the bazaar: Musings on Linux and open source by an accidental revolutionary* (Rev. ed.). Cambridge, MA: O'Reilly.

Resnick, D. (1997). Politics on the Internet: The normalization of cyberspace. *New Political Science*, 47–68.

Resnick, P., & Miller, J. (1996). PICS: Internet access controls without censorship. *Communications of the ACM, 39*(10), 87–93.

Samuelson, P. (1996). The copyright grab. *Wired*. Available at http://works.bepress .com/pamela_samuelson/240/

Schmidt, A. (2014). Hierarchies in networks: Emerging hybrids of networks and hierarchies for producing Internet security. In *Cyberspace and international relations* (pp. 181–202). Berlin, Germany: Springer.

Scholte, J. A. (2016). Process and power in Internet governance. Reflections on the IANA transition. *RIPE, Amsterdam*. Available at https://ripe72.ripe.net/presentations /20-ripe-72.pdf

Segal, A. (2016). *The hacked world order: How nations fight, trade, maneuver, and manipulate in the digital age*. New York, NY: PublicAffairs.

Selby, J. (2017). Data localization laws: Trade barriers or legitimate responses to cybersecurity risks, or both? *International Journal of Law and Information Technology, 25*(3), 213–232.

Sell, S. K. (2013). Revenge of the "nerds": Collective action against intellectual property maximalism in the global information age. *International Studies Review, 15*(1), 67–85.

Sørenson, E., & Torfing, J. (Eds.). (2007). *Theories of democratic network governance*. Basingstoke, UK: Palgrave Macmillan.

Souter, D. (2007). Whose summit? Whose information society? Developing countries and civil society at the world summit on the information society. Association for Progressive Communications. Retrieved from http://www.apc.org/en/pubs/manuals /governance/all/whose-summit-whose-information-society

Stadnik, I. (2019). Sovereign RUnet: What does it mean? Internet Governance Project, Georgia Institute of Technology. Retrieved from https://www.internetgovernance .org/research/sovereign-runet-what-does-it-mean/

Stallman, R. (2002). Free software, free society: Selected essays of Richard M. Stallman. Boston, MA: Free Software Foundation.

Trottier, D. (2016). Social media as surveillance: Rethinking visibility in a converging world. London, UK: Routledge.

UNDP. (1997). Human Development Report, 1997. New York, NY: United Nations Development Program.

van Eeten, M. J. (2017). Patching security governance: An empirical view of emergent governance mechanisms for cybersecurity. *Digital Policy, Regulation and Governance, 19*, 6.

van Eeten, M. J., and Mueller, M. (2013). Where is the governance in Internet governance? *New Media & Society, 15*(5), 720–736.

Wagner, B. (2016). *Global free expression—governing the boundaries of Internet content.* Switzerland: Springer International Publishing.

Weber, S. (2004). *The success of open source.* Cambridge, MA: Harvard University Press.

Weinberg, J. (2000). ICANN and the problem of legitimacy. *Duke Law Journal, 50,* 187.

Weinberg, J. (2001). Geeks and Greeks, *Info, 3*(4) 313–332.

Weinberg, J. (2011). Governments, privatization, and privatization: ICANN and the GAC. *Michigan Telecommunications and Technology Law Review, 18,* 1.

Wolfgarten, S. (2005). *Investigating large-scale Internet content filtering.* (Unpublished master's thesis). Dublin City University, Ireland. Retrieved from http://citeseerx.ist .psu.edu/viewdoc/download?doi=10.1.1.133.5778&rep=rep1&type=pdf

Wu, T. S. (1997). Cyberspace sovereignty? The Internet and the international system. *Harvard Journal of Law and Technology, 10*(3), 647–666.

Zuboff, S. (2015). Big other: Surveillance capitalism and the prospects of an information civilization. *Journal of Information Technology, 30*(1), 75–89.

Zuboff, S. (2019). *The age of surveillance capitalism: The fight for the future at the new frontier of power.* New York, NY: Profile Books.

4   Science and Technology Studies Approaches to Internet Governance: Controversies and Infrastructures as Internet Politics

Francesca Musiani

Research seeking to bridge Internet governance research with approaches in science and technology studies (STS) began growing in the second decade of the 2000s. Complementary to predominantly institutional approaches that set the agenda for Internet governance research in its early days—and are still prominently featured in it—STS approaches consider the agency of technology designers, policy makers, and users as they interact, in a distributed fashion, with technologies, rules, and regulations, leading to consequences with systemic effects that may, at times, be unintended. Social and political ordering is understood as a set of ongoing and contested processes—an ensemble of mundane practices that contribute to maintaining, hacking, circumventing, developing, testing, or using the Internet. Thus, conceptually, STS-informed Internet governance research relies on understanding governance as a normative system of systems, and it acknowledges the agency, often discreet yet pervasive, of both human and nonhuman actors and infrastructures. This chapter provides an overview of the current ways in which STS approaches are being applied to Internet governance research, and in particular, it focuses on controversy studies and infrastructure studies as two subsets of conceptual and methodological tools that are gaining increasing traction.

The chapter opens by retracing how STS first approached the Internet as a subject of study and examining how some key concepts in STS have found their way into Internet studies. It then proceeds to discuss the key aspects of applying an STS-informed analytical and empirical framework to Internet governance research, including assemblages and hybrid arrangements as means of "ordering" in Internet governance and the structural and performative effects of controversies on norm-making and decision-making.

In its second part, the chapter addresses more closely how STS-informed approaches to Internet governance analyze the structuring and performative effects of controversies on governance. It considers how controversies around claims of doing Internet governance (Epstein, Katzenbach, and Musiani 2016), made by different actors or groups, contribute to the creation of different worlds in which specific notions of governance make sense. The chapter will discuss how the study of controversies unpacks governance as a theoretical and operational concept by exposing the plurality of notions it refers to and their potential conflicts (Cheniti 2009a; Ziewitz and Pentzold 2014).

Finally, the chapter discusses how STS approaches to Internet governance focus on the agency of nonhuman actors and infrastructures as loci of governance mediation, such as information intermediaries, critical Internet resources, Internet exchange points, and surveillance and security devices (Musiani et al. 2016). The chapter will address how Internet governance takes shape through a myriad of technical architectures and infrastructures, "control points" (DeNardis 2014, 11) that are often discreet and invisible yet nevertheless crucial in building the increasingly public and articulate network of networks.

The chapter concludes by assessing the applicability of the STS lens to the study of Internet governance and its intersections with other approaches to Internet governance studies.

STS Meet Internet Studies: Introducing *Dispositifs* and Boundary Objects in the Study of the Internet

As Sandra Braman highlights in chapter 2, it may now be almost an academic given that the Internet is socio-technical in nature; however, it is often useful to recall this, as it reminds us that all processes carried out on and by means of the Internet have both a technical and a human component—including governance approaches. Indeed, as seminal articles such as DiMaggio et al.'s (2001) at the turn of the millennium have made explicit, the Internet as a research subject has "social implications" that make its analysis thoroughly relevant for social scientists.

Approaches issued from STS are among the tools that social science researchers have mobilized to investigate the complex nexus of Internet and society. Janet Abbate notes that

> STS can be useful to address the complex links between Internet technology and culture, which have blurred the frontiers of traditional categories. One of STS' tenets is to "open the black box" of technology to understand its functioning,

and understand how social relations and aims translate into artifacts. STS similarly offer models to describe how human and non-human actors exert joint agency in mediated environments. (2012, 170)

The STS turn in research addressing information and communication technologies is founded in the sociology of innovation's interest in technical objects and in social practices of appropriation of emerging technologies, in media sociology's increasing interest in information and communication technologies, and in the development of information and communication sciences: information and communication technologies become "interactional artifacts" (de Fornel 1994, 126). With the advent of the Internet, an object-centered, interdisciplinary field of study follows at the end of the 1990s, with the creation of Internet studies. The "STS turn" within this field calls for a particular attention paid to context and situated practices, the unveiling of the invisible work of Internet innovation, to paraphrase Susan Leigh Star (1999). STS approaches put emphasis on the practices that shape the management and governance of the Internet and its uses as a living reality, and they determine the ways in which it operates, works, resists, and functions. Furthermore, STS approaches invite consideration of values and rationalities of Internet and web practitioners not as *indicators* of how they perceive the world but as *resources and categories* that they deploy in specific circumstances in order to create and uphold specific configurations—in short, to actively organize their world (Cheniti 2009a).

STS have helped recognize technical artifacts' status as mediators inasmuch as they can modify the performativity of social actions. In this conception, it makes less sense to consider discourses and objects as separate spheres and more sense to understand discourses as circulating within objects, both spheres coconstructing each other (Gillespie, Boczkowski, and Foot 2014).

The notion of *dispositif* (in the Foucaldian sense, often translated as "device" in English) and boundary object are among the notable concepts at the intersection of Internet studies and STS. A socio-technical *dispositif* is defined as an assemblage of human and nonhuman actors, whose competencies and performances are distributed and whose existence is enabled by the workings of innovation. Moreover, the notion allows integration of agency (Proulx 2009) in the analysis for a more fine-grained appreciation of its collective dimension. With the concept of boundary object, Susan Leigh Star and James Griesemer (1989) have sought to analytically describe those processes in which actors coming from different social worlds, and called on to cooperate, manage to coordinate despite their diverging points of view. Because they account for the processes of delegation of work or other activities, or for the performative action of artifacts in the production of knowledge, boundary objects allow us to conceptualize the work of coordination, alignment, alliance, and translation among the different actors and the worlds they mobilize (Trompette and Vinck 2010).

These notions—which are both concepts and practical methodological tools—have been recognized as useful, alongside approaches in political science, international law, and economics, for tackling the macro questions of politics and power related to Internet governance by unpacking the micro practices of governance as mechanisms of distributed, semiformal, or reflexive coordination (Hofmann, Katzenbach, and Gollatz 2016), private ordering (Elkin-Koren 2012), and use of Internet resources.

STS Meet Internet Governance

Devices and boundary objects are just two among the concepts and tools that STS scholars have developed to study social order, an "effect generated by heterogeneous means" (Law 1992, 382), making the actual, continuous *processes* of ordering—of economic, political, discursive, technical, or other nature—the main focus of scientific inquiry. In this context, governance is understood as a set of dynamics of social ordering, which does not happen exclusively (and, possibly, not primarily) in politically designed institutions but is also enacted through mundane practices of people engaged in maintaining or challenging the social order (Flyverbom 2016; Woolgar and Neyland 2013). This approach to the study of social order implies new ways to question and reassemble what we think of as both the *Internet* and *governance*. Indeed, this sensibility for social order as continuous and contested processes translates into a growing attention to the mundane practices of all those involved in providing and maintaining, hacking and undermining, developing and testing, or simply using the network of networks (Musiani 2015), thus expanding the notion of governance in Internet governance.[1] These diverse practices are seen not as mere objects of regulation but as elements constitutive to articulating, reifying, and challenging established, emerging, or contested norms—it is the "doing" of Internet governance (Epstein, Katzenbach, and Musiani 2016).

It is also argued that an STS lens relieves the pressure of pursuing a single precise definition or perimeter of Internet governance as a mandatory prerequisite to any meaningful enquiry (Ziewitz and Pentzold 2014). Instead, STS approaches mostly consider the assumptions deriving from this operation as deterring the understanding of how Internet governance is enacted, in pervasive, networked, and often invisible ways.

Conceptually, STS-informed Internet governance research relies on understanding it as a normative system of systems, and it acknowledges the agency, often discreet and pervasive, of both human and nonhuman actors and infrastructures. Empirically, STS-informed Internet governance research focuses on the dynamics of ordering of assemblages and hybrid arrangements of Internet governance; on the structural and performative effects of controversies and destabilizations on norm making and decision-making or on the construction of authority and trust; and finally, on hybrid forums, private arrangements, and users and their practices. All these components help flesh out the doing of Internet governance and may be of use in revisiting central, yet ill-defined, concepts such as multistakeholderism. Epstein, Katzenbach, and Musiani (2016) unpack those key aspects, a discussion that I reprise and expand on here.

First, STS approaches acknowledge that technical and political governance are becoming increasingly intertwined. Scholars of Internet governance, and not only from STS, acknowledge more and more broadly the plurality of these modes of governance (two notable examples in this book are chapter 2 by Braman and chapter 6 by Hall and her coauthors); the next step is to incorporate their inability to be fully separated. STS approaches plead for an understanding of Internet governance as coexistence of different types of norms, elaborated in a variety of partially juxtaposed forums, and enforced, implemented, or merely suggested via a plurality of normative systems: law, technology, markets, discourses, and practices (Brousseau, Marzouki, and Méadel 2012). In chapter 10, Ten Oever and colleagues provide a useful overview of the mosaic that constitutes this multiplicity of normative sources and rightly point out that some normative features that had not been technically built into the Internet in its early days (because, of course, several of its spectacular developments could not be foreseen) subsequently had to be retroengineered in the network of networks or compensated for by other normative sources.

Acknowledging these diverse origins of norms relevant for the use and design of the Internet, most STS-informed Internet governance researchers base their understanding of governance in ordering instead of regulation, management, or control (Elkin-Koren 2012; Flyverbom 2011). According to them, the concept of ordering not only captures the normative effect of mundane practices and daily routines; it is also considered particularly well suited to the analysis of the organizational forms of global politics not as static entities but as assemblages—hybrid configurations constantly reshaping their purposes and procedures so as to connect and mobilize objects, subjects, and other elements around particular issues. In this light, the very institutions of Internet governance can also be explored with an STS-informed toolbox, as Mikkel Flyverbom (2011) has done for the United Nations, seeking to capture the complexity of global political governance arrangements as embedded practices.

An important feature of STS approaches is the investigation of nonhuman actors and infrastructures as loci of mediation. Indeed, information intermediaries, critical Internet resources, Internet exchange points, and surveillance and security devices play a crucial governance role alongside political, national and supranational institutions, and civil society organizations (Musiani et al. 2016). Internet governance takes shape through a myriad of infrastructures, devices, data fluxes and technical architectures that are often in the backstage, yet crucial in building the increasingly public and articulate network of networks. Laura DeNardis (2014, 11) defines these entities as infrastructural "control points," around which are entangled matters of technical and economic efficiency, as well as negotiations over human and societal values such as intellectual property rights, privacy, security, transparency.

Scholarly and policy discussions on "governing algorithms" and accountability of algorithms connect with this aspect and explore the governing power of algorithms (Mager 2012; Ziewitz 2016) as they predict and personalize users' behavior on the Internet and the perception other actors (institutions, firms) have of them. In line with this approach, and a little less recently, STS contributions have brought an important contribution to the study of the privatization of Internet governance (which Laura DeNardis has explored in several scholarly contributions, the latest in chapter 1)—that is, how decisions and actions that apply to governance are increasingly taken by private entities, in particular by the handful of giant tech companies such as Google and Facebook that, because of their size and quasi-monopolistic status, are in the position of setting de facto standards in policy-related arenas. As Rikke Frank Jørgensen explains in chapter 8, the privatization of Internet governance also poses methodological challenges for the (STS) researcher, due to the heavy dimension of industrial secrecy surrounding the activities of tech giants. Furthermore, as Hall and colleagues underline in chapter 6, the omnipresence, pervasiveness, and sheer amount of data produced and available in digital form, and the multiplicity of methods at the disposal of Internet companies to make sense of it, add further implications to the privatization of Internet governance in terms of informational asymmetry, privacy challenges, and surveillance.

Another way in which STS approaches add to institutional perspectives on Internet governance is the acknowledgment of the central role of invisible, mundane, and taken-for-granted practices in the constitution of design, regulation, and use of technology. It calls attention to, for example, acts of individuals in articulating Internet standards and to how instability was built into the early Internet so as to ensure the possibility of constant change (Braman 2016), the social aspects of crafting and enacting Internet-related policy, and how nontraditional forms of participation in discourse about Internet governance issues (i.e., multistakeholderism) become institutionalized (Epstein 2013). This part of STS approaches suggests that governance of the Internet, as a socio-technical system, is a social dynamic as well as a political one.

STS-informed approaches to Internet governance also address the structuring and performative effects of controversies on governance. Most prominently, they analyze how controversies around claims, made by different actors or groups, about "doing Internet governance" actually contribute to the creation of different worlds where specific notions of governance make sense (Epstein, Katzenbach, and Musiani 2016). Thus, the study of controversies unpacks governance as a theoretical and operational concept by exposing the plurality of notions it refers to and the consequences of their being in conflict (Cheniti 2009a; Ziewitz and Pentzold 2014). The very processes by which norms are created, renegotiated, put to the test, and realigned and raise conflicts are as crucial—and perhaps more crucial—in STS perspectives as the stabilized norms themselves (De Filippi and Loveluck 2016; Musiani, Mallard, and Méadel 2018).

As hinted earlier in the chapter, several concepts brought in by the STS toolbox can help unveil a number of situated practices on, by, and for the Internet that arguably constitute a vital part of doing Internet governance. For example, understanding Internet governance through the lens of Callon, Lascoumes, and Barthe's hybrid forums (2009)—entities meant to transform controversies into productive dialogue and bring about democracy—can enrich and revisit the concept of multistakeholderism (Malcolm 2008) by putting emphasis on actors' positioning and their evolving relationships to one another. The technology-embedded nature of most types of private sector interventions in Internet governance can be brought to the foreground by STS methods. Examining the relationship of Internet users with their devices and the values they embed does governance inasmuch as it reflects their commitment to a set of norms and to a community (Elkin-Koren 2012).

The final part of this chapter addresses in more detail two of these key aspects that, given the current social, economic, and political context in which the Internet as a device-of-devices is placed, we consider as particularly important and gaining traction among the lenses on Internet governance. The first is the structuring and performative effect of controversies on governance; the second is the agency of nonhuman actors and infrastructures as loci where governance is enacted and mediated.

Controversies and Their Performative Role in Internet Governance

Since the very early days of the Internet, being on and managing the network of networks has been about exercising control over particular functions of it that provide certain actors with the power and opportunity to act to their advantage; on the other hand, there is very rarely a single way to implement these functions or a single actor capable of controlling them. Thus, the Internet is controversial and contested, both target and instrument of governance, object of interest of a myriad of actors: from the most powerful and centralized to the average Internet user. Infrastructure can be understood as a fundamental place to exercise economic and political power (DeNardis and Musiani 2016, 3). Exposing these manifestations of power, which is often implicit and overlooked, is crucial in revealing conflicts and controversies about what an infrastructure is, who can benefit from it, or who can challenge it (Bernards and Campbell-Verduyn 2019).

Studying Internet-governance-related controversies (Pinch and Leuenberger 2006) is becoming one of the principal STS-informed ways to unpack Internet governance. Indeed, the Internet exhibits an increasing amount of contestation, in areas such as the interconnection agreements between Internet service providers (Meier-Hahn 2015), the debate around net neutrality (Marsden 2017), the use of deep packet inspection (Mueller, Kuehn, and Santoso 2012), the deployment of content filtering technologies

(Deibert and Crete-Nishihata 2012), ubiquitous surveillance measures, and the use of the DNS for regulatory aims (DeNardis and Hackl 2015). Furthermore, contentious politics, activism, and citizen-led protests are often embedded in the Internet and its applications, illustrated, for example, by the work of Milan and Ten Oever (2017) on civil society engagement within ICANN so as to encode human rights into Internet infrastructure or Ksenia Ermoshina's (2016) research on the shaping and use of citizen- and activist-oriented mobile and web applications and how the design of these tools shapes citizen participation and citizen-state interaction.

To put it in the words of Ziewitz and Pentzold (2010), controversies unveil different versions, according to different actors, of the worlds where notions of governance take place. Thus, for the analyst, the negotiations and controversies that take place around claims of "Internet governancing" (Cheniti 2009b) can be viewed as performative, inasmuch as they "both implicate and are implicated in creating the worlds in which a mode of governance makes sense" (Ziewitz and Pentzold 2010, 20).

Internet governance is particularly suited to all kinds of exit strategies and evolutions in power balances; thus, consensus-based modes of regulation become central, as norms cannot be totally binding and are permanently negotiated and challenged (Brousseau, Marzouki, and Méadel 2012, 35). As a consequence, the very processes by which norms evolve—put to the test and made the subject of conflict and realignment, destabilization and restabilization—become central because they provide different types of guarantees to the various stakeholders. Digital technologies themselves play a key role in this legitimation process because they become not only facilitators but guarantors of fairness and neutrality in controversial moments of these processes, as illustrated by research on the blockchain subtending Bitcoin (Musiani et al. 2018). In doing so, they also perform trust, both by automating procedures and by keeping track of all actions.

Contrary to what some institutional approaches may suggest, controversy, unsettling, destabilization, and restabilization are important parts of Internet governance institutions as well. For example, as Flyverbom (2011) shows, the Internet Governance Forum and other Internet governance organizational arrangements would not have been born without ample reconfigurations of two UN-linked entities (the Working Group on Internet Governance and Information and Communication Technologies Task Force). If examined through an STS lens, institutions show their ability to renegotiate and reconfigure themselves in moments of controversy in order to maintain momentum and, ultimately, authority. If not analyzed in such a light, the authority of Internet governance institutions would "otherwise come across as faits accomplis" (Flyverbom 2011, 6). Furthermore, as Julia Pohle (2016) argues, by focusing the analysis on actors' positionings and negotiations, and on processes rather than outcomes, it is possible to shed light on the contribution of multistakeholder processes and the validity of their results, albeit in the absence of binding outcomes.

Internet governance controversies and battles happen most of the time over "control points," as illustrated by Laura DeNardis (2014, 11). These control points range from the deepest layers of Internet infrastructure to the "last mile" of user access to the network, from the blocking of financial flows to the deliberate "kill-switches" of Internet-based services, from the "graduated response" termination of domestic Internet access to the attempted use of the DNS for copyright enforcement purposes, and from the Internet's backbone infrastructure to the establishment of interconnection agreements. They also include the de facto public policy role assumed by private information intermediaries, in the many ways they gather, collect, aggregate, select, and present data to users and to other actors of the Internet value chain—thereby enacting governance over privacy, freedom of expression, cultural diversity, and reputation (DeNardis 2012, 2014).

Infrastructure as Enacting and Mediating Governance

The term "infrastructure," as Sandra Braman's chapter 2 notes, is a potentially all-encompassing term that may be excessively vague without a definition. It commonly refers to the collective equipment necessary to human organization and activity, such as buildings, roads, bridges, and communications networks—in short, fully material and concrete artifacts. However, when it comes to the Internet (and its governance), Geoffrey Bowker and colleagues note that "beyond bricks, mortar, pipes or wires, infrastructure also encompasses more abstract entities, such as protocols (human and computer), standards, and memory," as well as "digital facilities and services … [such as] computational services, help desks, and data repositories to name a few" (Bowker et al. 2010, 97–98).

According to a body of work that draws inspiration from the research of Bowker and that of his colleagues, in particular Susan Leigh Star, the infrastructural quality of the network of networks is relational and conditional; infrastructures can be more usefully understood in terms of function than form. Thus, beyond objects whose infrastructural aspect is immediately obvious, such as bridges or pipes, a number of artifacts and entities that populate and shape the network of networks could be described as infrastructure because they have an infrastructural *function*—because they help structure, shape, enable, or constrain our being together on and with the Internet. In this sense, Internet infrastructures include physical objects—for example, submarine cables that carry global telecommunications, data centers that host digital content, and objects that are a priori much less concrete, such as the Internet protocol that allows the blockchain underlying Bitcoin to work.

A whole tradition of STS have explored the social and organizational dimensions of infrastructure of information and communication technologies, intended, thus, in these multiple senses of not only the purely material artifacts but also their logistical substrata. In particular, STS scholars have highlighted features that are of interest when studying complex socio-technical systems—for example, that infrastructure typically exists in the background, is invisible, and is frequently taken for granted (Star and Ruhleder 1994). Thus, it is argued, the politics inscribed in infrastructure by means of design and technical encodings is similarly difficult to trace. Yet it is an important task because the design of the "plumbing" of the Internet (Musiani 2012), the underlying practices, uses, and exchanges in a networked system, informs its adoption and (re)appropriation by users, its regulation, and its organizational forms. Several bodies of work, crossing Internet studies with the branch of STS called infrastructure studies, have sought to explore the social and organizational qualities of infrastructures subtending information networks and to find the materiality in the virtual of software and code (Blanchette 2011; Fuller 2008; Marino 2006). New concepts to account for the agency of infrastructure have been proposed, such as Annalisa Pelizza's "vectorial glance" (2016), which explores how interoperability of information systems, as a performative process of boundary reordering, redistributes authority and accountability: the small technical operations of interoperability projects become strategic sites where institutional shifts become visible.

STS-informed perspectives examining infrastructures have proliferated, but they at first received comparatively little attention from scholars of Internet *governance*, the pioneer in this regard having been Laura DeNardis with her article "Hidden Levers of Internet Control" (2012). They are now an important part of Internet governance scholarship, as contributions by Braman (2016) and Malcic (2016), on the work of the Internet's early designers, and De Filippi and Loveluck (2016), on the mixed technical and social governance subtending Bitcoin, have shown. In these contributions, Internet governance is understood as a set of socio-technical processes of innovation, digitalization, regulation, mobilization, co-optation, and circumvention.

Furthermore, contributions drawing from STS approaches in recent years have recognized not only that administrative and coordinating functions related to Internet infrastructure have always been instruments of power (DeNardis 2009) but that points of infrastructural control, regardless of their originally intended function, can serve as proxies to regain (or gain) control of or manipulate the flow of money, information, and the marketplace of ideas in the digital sphere—a phenomenon that has been called the "turn to infrastructure in Internet governance" (Musiani et al. 2016). This body of work addresses, for example, the use of the DNS as a tool for intellectual property rights enforcement (Merrill 2016) or information intermediaries' discretionary power to set their infrastructural practices to prioritize strategic interests over privacy commitments (Sargsyan 2016). Put together, these contributions show a shift from a "values-in-design" approach (Flanagan, Howe, and Nissenbaum 2008) to a politicization of Internet governance infrastructures (DeNardis 2009). That is to say, while values have entered the design of infrastructure for a long time, these values have been incorporated into technological infrastructure mostly to carry out its core functions; instead, the use of Internet infrastructure to carry out functions other than their intended objective can lead to important collateral damage for the stability and security of the Internet and the protection of online civil liberties (DeNardis and Musiani 2016). STS approaches, with their attention to situated practice and infrastructural agency, are well suited to bring these aspects to the foreground.

What Can STS Approaches Do for Internet Governance Research?

STS-informed analyses of the construction, materiality, and controversial potential of digital infrastructures offer new insights into the scope and limits of Internet-related technological change and its governance potential for at least two reasons.

First, the emphasis on the infrastructural nature of socio-technical systems makes it possible to shed new light on how the applications of alternative and emerging technologies mingle with dominant actors, objects, and processes. Infrastructure, as Star (1999, 382) puts it, does not develop from scratch but struggles with the inertia of what is already stabilized and inherits its strengths and limitations. Approaches to Internet infrastructure and its political weight also help explicitly highlight the contested and relational nature of technological change. New technologies do not change Internet operations on their own. On the contrary, technological change is mediated through fundamentally political struggles over the functioning and nature of the systems required to perform these infrastructural functions.

Second, highlighting efforts to position networked digital systems as material and infrastructural invites us to consider the contradictions of technological change. The denaturalization of socio-technical systems, making them black boxes, draws attention to these systems being flawed and subject to the possibility of failure. Star points out how the normally invisible quality of infrastructures becomes visible when they collapse (1999, 382). We are, for example, much more likely to notice our dependence on the electricity grid during a power cut than when everything is working as normal. The same could be said of systems that connect individuals, allow them to connect to the broadband Internet, convert purely digital addresses into addresses more intelligible for the human brain, or shape the blockchain.

As manifestations of failure, however, the material and process failures that underlie socio-technical systems are not only relevant in the instability that unmasks them. On the contrary, they are always important. Boundaries contributing to "infrastructure inequality" (Nelms 2016, 511) can help bring forward broader issues of access and can problematize information that can be standardized and operationalized and that which cannot. In a nutshell, it is by analyzing the politics of technological infrastructures and basing them in their materiality that we avoid implicitly and explicitly fetishizing the novelty of new technologies and develop a more nuanced perspective to understand what ultimately constitutes questionable—and, indeed, contentious—models of continuity and change.

If there is but one insight about Internet governance research to take away from this chapter, it is that, when examined through the STS lens, the Internet is not a given, static technological entity in need of regulation; it is the ensemble of technological elements of the network of networks and the different actors doing things with it that constitute, perpetuate, and contest sociopolitical order. In addition to the technical decisions about the design and operation of the network, formal law and regulation, and the forces of the market, a number of rather mundane and taken-for-granted activities, driven by heterogeneous and often competing visions or based on inherently social and political arrangements of trust and consensus, contribute to Internet governance as it is today.

Conclusion

As the Internet more and more becomes humanity's primary global facility, marketplace, and public sphere, sociopolitical and socio-technical controversies become an increasingly important part of what lies under the Internet governance label. The STS toolbox provides one of the most interesting opportunities for them to be thoroughly accounted for, richly described, and extensively analyzed, with notions that are both *concepts* (they suggest a vision of how the world goes, what drives its operations, and what makes them meaningful in political terms) and *methods* (each of these notions is also a practical way to apprehend the inner workings of Internet governance on the field). In this sense, recent research seeking to merge STS and Internet governance is indeed a blueprint for a controversy-based and infrastructure-based understanding of the backstage of today's Internet politics.

STS methods come, of course, with their own set of challenges. Looking at the mundane, the "shaping invisible" (Musiani 2018), which usually escapes the public radar—often even the scholarly radar—implies identifying the right terrain, singling out the ways to get to it, and finally patiently negotiating access to it (see Jørgensen's chapter 8), because an in-depth ethnographic work is a necessary precondition to a meaningful STS endeavor. This negotiation is, sometimes, not with potential fieldwork actors but with the researcher's own set of competences: to analyze and, more, to clearly *analytically describe* environments requires a high level of technicality that first needs to be mastered by the researcher. Thus, STS researchers of Internet governance (as is the case for many scholars of the governance of other technical systems and devices) often have composite disciplinary backgrounds, having arrived to STS methods only after previous training in computer science and engineering.

The choice of which fieldwork to address in depth—usually one or a few case studies—brings with it questions of criteria selection for that choice and of the representativeness of the selected cases. And finally, closely related to the previous point—despite the difficulty of generalizing to broader principles, which is intrinsic to the STS approach—for their work to be meaningful in a broader dialogue with other disciplines, STS-inspired Internet governance researchers should guard against falling into a common trap of their discipline of disciplines: making the language of complexity and heterogeneity the main protagonist of their analyses to the point of clouding conclusions behind it.

The increasing attention dedicated by STS scholars to the Internet governance field has not, of course, grown in isolation. In addition to the lineage of Internet studies introduced earlier in the chapter, a significant body of existing STS literature provides insights into distributed participation in techno-scientific controversies, and Internet governance research can learn from governance of and *by* science and technology in other contemporary, complex socio-technical domains such as environment, health, nanotechnologies, and genetic engineering (see, e.g., Irwin 2006). Similarly, Internet governance research in other, more historical disciplines, mainly focused on the institutional level and the role of the state—political science, law, history, international relations, and institutional economics—can speak to STS and help, for example, mitigate some of the undesired consequences of STS approaches described earlier.

The nexus between STS-inspired Internet governance studies and other bodies of STS on infrastructures and socio-technical controversies is likely to become even more inextricably entangled as the reach of Internet infrastructure actors extends to other types of infrastructures. Larry Page once predicted that "Google would be building airports and cities," and the Internet giants are readying to extend themselves: while it has long been believed that the influence of digital actors would remain confined to software, dematerialized content, and information, it starts to be clear that they are using their mastery in these areas to take positions in non-digital markets, be it transport, infrastructure management, or banking. Google may not be building cities yet, but directly or through its investments it is already playing a role as a mobility organizer. IBM participates in the management of water supply infrastructure in several cities. With the connection of infrastructures and objects, the organization of physical flows requires the control of information flows. Massive data are at the heart of this movement, which calls into question the positions of the historical players in these markets. Internet governance and its study in the near future should take into account these ongoing developments about the perimeter and the very nature of Internet infrastructure, just as, in the recent past, they started to acknowledge the shift from governance *of* Internet infrastructure to governance *by* infrastructure.

The STS focus on unpacking some of the mundane elements of the Internet—such as technical details and minutiae, invisible maintenance work, specific case studies, and close follow-ups of controversies—is a necessary complement, not a substitute, to those efforts that seek to elaborate general principles and theories about the ways in which the distribution of power and resources, in short, the world of politics and governance, works.

Acknowledgments

This chapter summarizes and updates work first outlined in Musiani (2015), further developed in Musiani (2018), and conducted jointly with several colleagues (DeNardis and Musiani 2016; Epstein, Katzenbach, and Musiani 2016; Musiani and Méadel 2016; Musiani, Pelizza, and Milan 2016; Musiani and Schafer 2019). My heartfelt thanks to Laura DeNardis, Dmitry Epstein, Christian Katzenbach, Cécile Méadel, Stefania Milan, Annalisa Pelizza, and Valérie Schafer, without whom this chapter would not exist.

During the writing of this chapter, I was supported by the European Union's Horizon 2020 Framework Programme for Research and Innovation (H2020-ICT-2015, ICT-10–2015) under grant agreement number 688722–NEXTLEAP, and I am currently supported by the French National Agency for Research (ANR), under grant ANR-17-CE26-0020–ResisTIC.

Note

1. A debate is ongoing among STS Internet governance scholars on whether user agency and practice should be included in Internet governance, Laura DeNardis (2014) notably saying it should not be included. Interestingly, as Mueller and Badiei in chapter 3 show, this uncertainty about whether certain user practices should or should not be included in Internet governance may account for its underrepresentation as a research topic in sociology and social sciences journals, because several sociologists are in fact studying issues relevant to Internet governance without labeling it as such.

References

Abbate, J. (2012). L'histoire de l'Internet au prisme des STS. *Le temps des médias, 18*, 170–180.

Bernards, N., & Campbell-Verduyn, M. (2019). Understanding technological change in global finance through infrastructures: Introduction to Review of International

Political Economy Special Issue "The Changing Technological Infrastructures of Global Finance." *Review of International Political Economy, 26*(5), 773–789. doi.org/10.1080/09692290.2019.1625420

Blanchette, J.-F. (2011). A material history of bits. *Journal of the Association for Information Science and Technology, 62*(6), 1042–1057.

Bowker, G. C., Baker, K., Millerand, F., & Ribes, D. (2010). Toward information infrastructure studies: Ways of knowing in a networked environment. In J. Hunsinger, L. Klastrup, and M. Allen (Eds.), *International Handbook of Internet Research* (pp. 97–117). Dordrecht, Netherlands: Springer.

Braman, S. (2016). Instability and Internet design. *Internet Policy Review, 5*(3). doi:10.14763/2016.3.429

Brousseau, E., Marzouki, M., & Méadel, C. (Eds.). (2012). *Governance, regulation and powers on the Internet*. Cambridge, UK: Cambridge University Press.

Callon, M., Lascoumes, P., & Barthe, Y. (2009). *Acting in an uncertain world: An essay on technical democracy*. Cambridge, MA: MIT Press.

Cheniti, T. (2009a). *Global Internet governance in practice: Mundane encounters and multiple enactments* (Unpublished doctoral dissertation). University of Oxford, UK.

Cheniti, T. (2009b, March). *Internet governanc-ing*. Paper presented at the Modes of Governance in Digitally Networked Environments Workshop, Oxford Internet Institute, Oxford, UK.

De Filippi, P., & Loveluck, B. (2016). The invisible politics of Bitcoin: Governance crisis of a decentralized infrastructure. *Internet Policy Review, 5*(3). doi:10.14763/2016.3.427

de Fornel, M. (1994). Le cadre interactionnel de l'échange visiophonique. *Réseaux, 64*, 107–132.

Deibert, R. J., & Crete-Nishihata, M. (2012). Global governance and the spread of cyberspace controls. *Global Governance: A Review of Multilateralism and International Organizations, 18*(3), 339–361.

DeNardis, L. (2009). *Protocol politics: The globalization of Internet governance*. Cambridge, MA: MIT Press.

DeNardis, L. (2012). Hidden levers of Internet control: An infrastructure-based theory of Internet governance. *Journal of Information, Communication & Society, 15*(3), 1–19.

DeNardis, L. (2014). *The global war for Internet governance*. New Haven, CT: Yale University Press.

DeNardis, L., & Hackl, A. M. (2015). Internet governance by social media platforms. *Telecommunications Policy, 39*(9), 761–770. doi:10.1016/j.telpol.2015.04.003

DeNardis, L., & Musiani, F. (2016). Introduction: Governance by infrastructure. In F. Musiani, D. L. Cogburn, L. DeNardis, & N. S. Levinson (Eds.), *The turn to infrastructure in Internet governance* (pp. 3–21). New York, NY: Palgrave Macmillan.

DiMaggio, P., Hargittai, E., Neuman, W. R., & Robinson, J. P. (2001). Social implications of the Internet. *Annual Review of Sociology, 27*(1), 307–336.

Elkin-Koren, N. (2012). Governing access to user-generated content: The changing nature of private ordering in digital networks. In E. Brousseau, M. Marzouki, C. Méadel (Eds.), *Governance, regulations and powers on the Internet* (pp. 318–343). Cambridge, UK: Cambridge University Press.

Epstein, D. (2013). The making of institutions of information governance: The case of the Internet Governance Forum. *Journal of Information Technology, 28*(2), 137–149.

Epstein, D., Katzenbach, C., & Musiani, F. (2016). Doing Internet governance: Practices, controversies, infrastructures, and institutions. *Internet Policy Review, 5*(3). doi:10.14763/2016.3.435

Ermoshina, K. (2016). *Au code, citoyens: Mise en technologie de problèmes publics* (Unpublished doctoral dissertation). MINES ParisTech.

Flanagan, M., Howe, D. C., & Nissenbaum, H. (2008). Embodying values in technology: Theory and practice. In J. van den Hoven & J. Weckert (Eds.), *Information technology and moral philosophy* (pp. 322–353). Cambridge, UK: Cambridge University Press.

Flyverbom, M. (2011). *The power of networks: Organizing the global politics of the Internet*. Cheltenham, UK: Edward Elgar.

Flyverbom, M. (2016). Disclosing and concealing: Internet governance, information control and the management of visibility. *Internet Policy Review, 5*(3). doi:10.14763/2016.3.428

Fuller, M. (Ed.). (2008). *Software studies: A lexicon*. Cambridge, MA: MIT Press.

Gillespie, T., Boczkowski, P., & Foot, K. (Eds.). 2014). *Media technologies: Essays on communication, materiality and society*. Cambridge, MA: MIT Press.

Hofmann, J., Katzenbach, C., & Gollatz, K. (2016). Between coordination and regulation: Finding the governance in Internet governance. *New Media & Society, 19*(9), 1406–1423.

Irwin, A. (2006). The politics of talk: Coming to terms with the "new" scientific governance. *Social Studies of Science, 36*(2): 299–320.

Law, J. (1992). Notes on the theory of the actor-network: Ordering, strategy, and heterogeneity. *Systems Practice, 5*(4), 379–393.

Mager, A. (2012). Algorithmic ideology: How capitalist society shapes search engines. *Information, Communication & Society, 15*(5), 769–787.

Malcic, S. (2016). The problem of future users: How constructing the DNS shaped Internet governance. *Internet Policy Review, 5*(3). doi:10.14763/2016.3.434

Malcolm, J. (2008). *Multi-stakeholder governance and the Internet Governance Forum.* Wembley, Australia: Terminus Press.

Marino, M. C. (2006). Critical code studies. *Electronic Book Review.* Retrieved from http://www.electronicbookreview.com/thread/electropoetics/codology

Marsden, C. T. (2017). *Network neutrality: From policy to law to regulation.* Manchester, UK: Manchester University Press.

Meier-Hahn, U. (2015, February 5). Internet interconnection: Networking in uncertain terrain [Blog post]. *RIPE Labs.* Retrieved from https://labs.ripe.net/Members/uta_meier_hahn/internet-interconnection-networking-in-uncertain-terrain

Merrill, K. (2016). Domains of control: Governance of and by the domain name system. In F. Musiani, D. L. Cogburn, L. DeNardis, & N. S. Levinson (Eds.), *The turn to infrastructure in Internet governance* (pp. 89–106). New York, NY: Palgrave Macmillan.

Milan, S., & ten Oever, N. (2017). Coding and encoding rights in Internet infrastructure. *Internet Policy Review, 6*(1). doi:10.14763/2017.1.442

Mueller, M. L., Kuehn, A., & Santoso, S. M. (2012). Policing the network: Using DPI for copyright enforcement. *Surveillance & Society, 9*(4), 348–364.

Musiani, F. (2012). Caring about the plumbing: On the importance of architectures in social studies of (peer-to-peer) technology. *Journal of Peer Production, 1.* Retrieved from http://peerproduction.net/issues/issue-1/peer-reviewed-papers/caring-about-the-plumbing/

Musiani, F. (2015). Practice, plurality, performativity and plumbing: Internet governance research meets science and technology studies. *Science, Technology and Human Values, 40*(2): 272–286.

Musiani, F. (2018). L'invisible qui façonne. Etudes d'infrastructure et gouvernance d'Internet. *Tracés, 35,* 161–176.

Musiani, F., Cogburn, D. L., DeNardis, L., & Levinson, N. S. (Eds.). (2016). *The turn to infrastructure in Internet governance.* New York, NY: Palgrave Macmillan.

Musiani, F., Mallard, A., & Méadel, C. (2018). Governing what wasn't meant to be governed: A controversy-based approach to the study of governance in Bitcoin. In M. Campbell-Verduyn (Ed.), *Bitcoin and beyond* (pp. 133–156). London, UK: Routledge.

Musiani, F., & Méadel, C. (2016). "Reclaiming the Internet" with distributed architectures: An introduction. *First Monday, 21*(12). Retrieved from http://firstmonday.org/ojs/index.php/fm/article/view/7101

Musiani, F., Pelizza, A., & Milan, S. (2016). Materializing Governance by Infrastructure. *4S Preview Blog.* Retrieved from http://www.4sonline.org/blog/post/materializing_governance_by_infrastructure

Musiani, F., & Schafer, V. (2019). Science and technology studies approaches to web history. In N. Brügger & I. Milligan (Eds.), *Handbook of Web History* (pp. 73–85). Thousand Oaks, CA: Sage.

Nelms, T. C. (2016). Alt. economy: Strategies, tensions, challenges. *Journal of Cultural Economy, 9*(5), 507–512.

Pelizza, A. (2016). Developing the vectorial glance: Infrastructural inversion for the new agenda on governmental information systems. *Science, Technology and Human Values, 41*(2), 298–321.

Pinch, T., & Leuenberger, C. (2006). *Studying scientific controversy from the STS perspective.* Paper presented at the Science Controversy and Democracy Conference, Taipei, Taiwan.

Pohle, J. (2016). Multistakeholder governance processes as production sites: Enhanced cooperation "in the making." *Internet Policy Review, 5*(3). doi:10.14763/2016.3.432

Proulx, S. (2009, June). L'intelligence du grand nombre: la puissance d'agir des contributeurs sur Internet—limites et possibilités. *7ème colloque du chapitre français de l'ISKO, Intelligence collective et organisation des connaissances*, Lyon, France. http://pro.ovh.net/~iskofran/pdf/isko2009/PROULX.pdf

Sargsyan, T. (2016). The privacy role of information intermediaries through self-regulation. *Internet Policy Review, 5*(4). doi:10.14763/2016.4.438

Star, S. L. (1999). The ethnography of infrastructure. *American Behavioral Scientist, 43*(3), 377–391.

Star, S. L., & Griesemer, J. (1989). Institutional ecology, "translations" and boundary objects: Amateurs and professionals in Berkeley's Museum of Vertebrate Zoology, 1907–39. *Social Studies of Science, 19*(3), 387–420.

Star, S. L., & Ruhleder, K. (1994). Steps towards an ecology of infrastructure: Complex problems in design and access for large-scale collaborative systems. *Proceedings of the Conference on Computer Supported Cooperative Work* (pp. 253–264). Chapel Hill, NC: ACM Press.

Trompette, P., & Vinck, D. (Eds.). (2009). Retour sur la notion d'objet-frontière. *Revue d'anthropologie des connaissances, 3*(1), 5–27.

Woolgar, S., & Neyland, D. (2013). *Mundane governance: Ontology and accountability.* Oxford, UK: Oxford University Press.

Ziewitz, M. (2016). Governing algorithms: Myth, mess, and methods. *Science, Technology and Human Values, 41*(1), 3–16.

Ziewitz, M., and Pentzold, C. (2010, October). *Modes of Governance in Digitally Networked Environments: A Workshop Report* (Discussion Paper No. 19). Oxford, UK: Oxford Internet Institute Forum.

Ziewitz, M., and Pentzold, C. (2014). In search of Internet governance: Performing order in digitally networked environments. *New Media & Society, 16*(2), 306–322.

5   A Legal Lens into Internet Governance

Rolf H. Weber

Law as a Flexible System

In the field of Internet governance, the fast-changing technological and political environment challenges the suitability of traditional regulatory regimes; this assessment is justified even if the emphatic pronouncements in John Perry Barlow's (1996) *Declaration of the Independence of Cyberspace* have turned out to be not realistic. The dichotomy between the global reach of the infrastructure and the national embedment of the normative framework occasions the need for new legal research concepts and methods. Issues such as legitimacy, regulatory quality, technical standardization, and accountability are to be addressed (Weber 2016a). As I show in this chapter, legal scholars need to tackle these challenges.

Traditionally, the legal order was based on a communal, later on a national, normative framework that was complemented by self-regulatory instruments and, since the 19th century, partly by multilateral agreements. This regulatory framework, developed for the real world, is exposed to limitations if applied to the online world designed by the new information technologies. Therefore, the traditional understanding of political structures as command by a specific body that induces people to execute certain actions—in the sense that people think about what to choose and what to do—should be replaced in the Internet governance context by a more inclusive approach. Before discussing the details of such an approach, I outline some basic principles of legal theory and the relevant guiding regulatory strategies for Internet governance.

Structural and Open System

In legal theory, law is seen as a structural system being composed of an organized or connected group of objects (terms, units, or categories) forming a complex unity. Legal norms are usually expressed in a linguistic form and are designed to give guidance about the desired behavior (Weber 2002, 32). The addressees, be it the whole society or a concerned part thereof, are supposed to take note of the contents of law. Thereby, legal concepts help support adequate normative reasoning and stabilize normative expectations (Mahler 2014, 27–33; Weber 2014b, 33).

The functions of law crystallize in a system of rules and institutions that underpin civil society, facilitate orderly interaction, and resolve conflicts and disputes arising in spite of the rules (Chik 2010). The creation of law can happen by way of different processes—for example, negotiations among the concerned norm addressees (a "social contract," following the concept of Rousseau), imposition of legal rules by the governing body, or evolution in self-regulatory mechanisms (Weber 2014b, 33–34; see also Amstutz 2011, 395).

The legal system is not a predetermined construct—that is, the legal system is embedded in other socially relevant systems. Moreover, exchange and interchange between different social systems make the legal order porous. The open system approach is mainly influenced by the advances in cybernetics and information theory. In principle, the complexity of any system depends on the inclusion of other organized systems.[1] Since modern societies are differentiated into a plurality of subsystems, a framework of sociological "functionalism" must be developed (Weber 2014b, 47).

A "meaningful law" in an open system is composed of norms that are perceived as legally binding, thereby inducing the addressees to acknowledge the authority of the rulemaking body and to comply with the rules (Reed 2012, 70–73, 105, 106). As a result, law should be able to regulate behavior and to allow people in a community to determine the limits of what can and cannot be done in their collective interests (Weber 2014b, 34; see also Mahler 2019, 72–94; and Sandra Braman's chapter 2).

Relative Autonomy and Flexibility of Law

In view of the rapid technological developments that cause social changes, it is imperative to realize a flexible legal framework in an open society. This flexibility presupposes that the legal rules profit from a certain degree of legal autonomy notwithstanding the linkages between different subsystems in society.

The open society approach requires an assessment of the interdependence between normative concepts and other social sciences' perspectives.

In this context, legal theory scholars have defined criteria for the relative autonomy of law (see Post 1991, vii–viii): (1) "Autonomy" means that the law is not equal to and not fully dependent on other social sciences; (2) the word "relative" evidences that exchanges between the law and other social spheres take place in both directions (Weber 2002, 36–37).

The theoretical foundation of relative autonomy shows that other social sciences are not in a position to rule out any legal flexibility being of importance because law needs to be able to react to changing circumstances (Weber 2014b, 49–50). Nevertheless, the autonomy model does not directly find a clear distinction between law and no law. Generally, however, law may draw on insights from some other fields of discourse while retaining its separate entity.

Substance and Change of Law

Legal rules usually contain information having a guiding or even coercive effect on the members of civil society. The legal framework is composed of different instruments (for a general overview, see Weber 2002, 56):

- Multilateral or bilateral agreements binding the ratifying countries within the scope of the agreed regulations
- Fundamental norms stating substantial values and policies governing the life of the citizens and organizations in a country (usually the constitution)
- General rules applying to individuals and organizations in the form of a law or ordinance
- Specific decisions ruling on certain aspects of a legal relation

To avoid a legal system becoming rigid, mechanisms must be introduced that allow a change of the law according to social needs and circumstances. Notwithstanding that the predictability of the law requires a stable structure, the adaptability of legal rules, resilient to change, keeps the law intact in case of a relevant social change.[2]

However, before adapting existing laws, lawmakers should consider that legal changes are economically not without cost and do have a social impact: Laws are not created in a vacuum. New legal rules can be risky and costly. Addressees of legal rules may have a limited capacity for attention. New legal rules impose learning costs on the legal profession (see Weber 2002, 39).

The development of appropriate guidelines for potential changes of law is particularly important in the Internet field since the technological environment is fast evolving and the global reach of the infrastructure is inherent.

Regulatory Strategies

Regulatory strategies cannot be implemented without regard to the political landscape that is in the process of being established in the Internet governance field.

Political Visions of Rulemaking

Looking at the experience of the last few years, it seems obvious that the identification of the relevant political structures and their shortcomings as well as the assessment of the international legal order's potential merit greater attention (see Rioux 2013, 37, 49–54, for an overview of the constellations of regulatory instruments in global governance). The appropriateness of the legal framework encompassing Internet governance depends on the ability of the policy makers to embrace new approaches using different normative tools. Therefore, the implementation of legal instruments must be done with great care and prudence to avoid undesired effects (Weber 2014b, 102).

Two visions of political power can be distinguished: the dominance of the state power as founded on the sovereignty concept and the power distribution relying on various stakeholders (Klimburg 2013). The two competing models have an impact not only on the international rulemaking agenda but also on the design of supranational institutions and the role of sovereign states. Therefore, the decision for one of the two models influences the decision-making processes and, indirectly, the outcome of deliberations. An example can be seen in the different approaches pursued at the World Conference on International Telecommunications in Dubai in December 2012, organized by the International Telecommunication Union, and at the plenary conferences of the Internet Corporation for Assigned Names and Numbers (ICANN) (see Weber 2013, 95, 98, 101).

The current challenges in Internet governance regulation by nature require a broader and more collective decision-making than in a nation-state. In times of globalization, the movement toward global governance is unavoidable and the structure of international law will need some adaptations. The crucial point concerns the appropriate balance of power between sovereign states, nonstate actors such as businesses and individuals, and new geographic or functional entities in a power-sharing framework (DeNardis 2014, 23).

As a result, global governance must enshrine collective efforts enabling the concerned persons to identify, understand, and address the global problems that go beyond the capacity of individual states to solve. Political theory may operate at different levels: A global framework needs to be combined with domestic political theory in order to assess the necessary interplay of the different levels. A global political theory must be able to provide guidance as to what principles should be adopted and which principles should be implemented in reality. Rules are needed that help determine how general principles can be applied to specific issues (Weber 2014b, 105).

Quality of Regulation

An important aspect of Internet governance debates and its normative framework concerns the quality of regulation. Several criteria can improve the desired quality of regulation; the following questions should be taken into account (see Baldwin, Cave, and Lodge 2012, 27–33):

- Is the regulatory action supported by legislative authority?
- Does the regime implement an appropriate scheme of accountability?
- Are procedures fair, accessible, and open?
- Is the regulator acting with sufficient expertise?
- Can the regulatory regime be assessed as an efficient system?

In an attempt to improve regulatory quality, the Organization for Economic Cooperation and Development issued *Guiding Principles for Regulatory Quality and Performance* (2005), which encompasses an extended scope of relevant aspects that reflect the social and environmental developments:

- Adoption of broad programs of regulatory reform that establish key objectives and frameworks for implementation at the political level
- Systematic assessment of impacts and review of regulations to ensure that the intended objectives are efficiently and effectively reached in a changing and complex economic and social environment
- Assurance that regulations, regulatory institutions charged with their implementation, and regulatory processes are transparent and nondiscriminatory

- Elimination of unnecessary regulatory barriers to trade and investment through continued liberalization and enhancement of market openness throughout the regulatory processes
- Identification of important linkages with other policy objectives and development of policies to achieve a harmonized regime

In Internet governance, the appropriateness of these elements remains unchanged, but the approach needs to be widened.

Informal Rulemaking and New Regulatory Models

Experience has shown that the traditional legal instruments can hardly cope with the challenges of Internet governance. Therefore, rulemaking through informal social relations based on the human evolution from individuals into members of society must gain importance. The establishment of social structures has already been a line of thought, expressed by the philosophers of the 17th century (Thomas Hobbes, Jean-Jacques Rousseau). On the basis of the understanding of civil society, normative expectation encompassing substantive legal principles can evolve over time (Weber 2016a, 195, 201).

The discussions about Internet governance over the last 12 years have led to a dynamic regulatory matrix, or a hypercomplex structural match (Jørgensen 2013, 22–24). Such kinds of complex structures have been described as polycentric regulation involving different communities in the rulemaking processes (Murray 2007, 47, 234–235). If the participants in a polycentric regulation scheme have a shared set of normative beliefs, notions of validity of the rules and common policies grow; therefore, digitization is in a position for improving connections and facilitating the exchange of communications. However, the polycentric regulation concept's weakness consists in the practical problems of rulemaking pluralism and fragmentation. The Internet is in need of a (at least partially) coordinated set of rules; discretionary pluralism would destroy its value since incompatible legal rules could have a negative impact on its global reach (Weber 2016a, 202–203; for a general discussion of polycentric regulation, see Senn 2011, 31, 170).

Another theory is the concept of the so-called hybrid regulation; the term "hybrid" can be described as a combination of a contradictory difference, marked not by either-or but by both-and (Weitzenboeck 2014, 49, 62). A similar approach is the so-called mesh theory, which is based on the acknowledgment that a paradigm shift occurs with the profound transformation from a pyramid model of the government at the top to a network model (Ost and Van de Kerchove 2002, 14). Following this conceptual approach, a "network communitarianism" can evolve as a process of discourse and dialogue between the individual and the society (Murray 2007, 68). However, the regulatory legitimacy of these concepts can be challenged because substantial discretion for the assessment of the rule-making's quality is left open (see Weber 2016a, 203–204).

Another approach conceptualizing a world of hybrid legal spaces is the theory of global legal pluralism; this concept intends to encompass more than one legal or quasi-legal regime in the same field (Berman 2007, 1155–1158). The most recent theoretical approach to overcome the problems of previous regulatory concepts pleads for "global experimentalist governance," an institutionalized transnational process of participatory and multilevel problem-solving that frames critical issues in an open-ended way by subjecting them to periodic revisions (de Búrca, Keohane, and Sabel 2014, 477).

Notwithstanding several merits of the newest regulatory concepts, even an ideal model does not address important substantive principles in Internet governance—for example, legitimacy and multistakeholderism—therefore, these two issues are discussed in more detail in the following.

Legitimacy of the Internet Governance Legal Framework

Notion and Scope

The word "legitimacy" can be traced back to the Latin word "legitimus" as meaning "lawful, according to law." Legitimacy reflects an authority's right to rule and embraces the justification of ruling power giving the governed the feeling that their own values are represented in a decision-making context (Weber 2014b, 102–103).

Legitimacy in a wider sense can also encompass an ethical-philosophical dimension that puts legitimacy above positive law. A distinction is made between normative legitimacy theories, setting out general criteria for evaluating the right to rule, and empirical legitimacy theories, focusing on belief systems of those subject to government. As a result, legitimacy can be justified either by formal ideas as the rule of law rationale (legality) or by substantive value rationality based on morality and justice (see also Clark 2005, 18–19).

According to Jürgen Habermas (1992), a source-oriented perception qualifies an authority as legitimate if it refers to the demos, the public.

Procedural steps (or adequate procedures, in the terminology of Niklas Luhmann [1975]) within the different governing entities may enhance the legitimacy of policy-making decisions (see Weber 2009, 109, for a detailed analysis of the legitimacy of policy making from a theoretical perspective). Legitimacy also describes the relevant elements of governance that validate institutional decisions emanating from a right process.

Normative Principle

Legitimacy must be designed in line with constitutional values and principles. As architectural pillars, three concepts can be put in place: "legal, morality, and constitutionality," which are able to "mark out the terrain within which the practice of legitimacy tends to take place" (Clark 2005, 18). Legitimacy plays, then, the role of a reconciling norm, enabling consensus on how the three pillars are to be accommodated among each other (Clark 2005, 19; for the rule of law, see also Braman's chapter 2).

The assessment of legitimacy can also be done from the perspective of regulatory purposes and standards, regulatory instruments, regulatory effectiveness, and regulatory connection. These perspectives are gaining importance because legitimacy questions are becoming weightier not only for the international society in general but also for the stability of the international order (Weber 2014b, 113).

As mentioned, procedural elements are crucial for the acknowledgment of a right process. However, procedure must be complemented by a substantive conception that looks at the outcome of the legitimizing procedures (a result-oriented type of legitimacy). Such an approach depends on the values deemed as right by the stakeholders concerned, thus in part justifying them as legitimizing procedures. To avoid subjective perceptions of legitimate values prevailing, Habermas tried to link the procedural aspects with specific notions of contents. This "discourse principle" assumes that those norms can claim validity that receive the approval of all potentially affected persons, insofar as they participate in a free and rational discourse (Habermas 1992, 161).

Concretization for Internet Governance

In Internet governance, the implementation of appropriate organizational rules in the concerned social communities is a necessity. The actual process can choose between different avenues. On the one hand, moral norms falling under the notion of netiquette are relevant for online macrocommunities. On the other hand, the administration of the Internet, seen as a microcommunity, needs some basics of taxonomy (Weber 2014b, 112; for the history of the Internet, see Mueller and Badiei's chapter 3).

ICANN as the main organization in Internet governance is a private organization. However, over time its legitimacy increased, partly due to the less strong ties with the government of the United States, partly due to the increased participation by other stakeholders over the last years. Further improvements of legitimacy are obviously possible, particularly regarding legal remedies (for example, an independent mediation and arbitration system); in general, however, the acknowledgment of legitimacy by ICANN officials is on the right track.

From a theoretical perspective, the original self-regulatory mechanisms addressing legitimacy issues have moved to a more democratic and equally harder normative framework. Thus, the legal impact on governance elements has become stronger, and some quality criteria of regulation are fulfilled to a wider extent.

Multistakeholderism in Internet Governance

Notion and Fundamentals

Before the second World Summit on the Information Society in late 2005, the Working Group on Internet Governance (2005) introduced a widely accepted working definition of multistakeholderism. This definition refers to the "development and application by Governments, the private sector and civil society, in their respective roles, of shared principles, norms, rules, decision-making procedures, and programmes that shape the evolution and use of the Internet." Therefore, the interests of the stakeholders involved should be designed by participatory mechanisms reflecting the whole society's view. Multistakeholderism is not a completely new phenomenon evoked in the context of Internet governance; earlier developments concerned labor and sustainability fields (Weber 2016b, 247–249; see also DeNardis's chapter 1 and Hofmann's chapter 12).

Four fundamental questions are at stake: How do governance groups best match challenges with the organizations, experts, and networks? How can governing bodies and entities be most able to help develop legitimate, effective, and efficient solutions? How should the flow of information and knowledge necessary for a successful governance be structured? How can different governance groups approach coordination between geographically different governance networks to avoid conflicting interests? (see ICANN/WEF Panel on Global Internet Cooperation and Governance Mechanisms 2014).

Practical considerations lead to the following additional questions: How can greater transparency and dialogue between different civil society groups and standards experts be introduced? How can standards be developed rapidly with the scrutiny of the increasing multistakeholder arrangements? (Brown and Marsden 2013, 200).

Basic values of multistakeholder models are openness (access to discussions, negotiations, and decisions), transparency (clear formal and substantive regimes with appropriate representation), accessibility (for information sources and procedures), accountability (responsibility of decision makers), credibility (general acceptance of decision makers), and consensus-orientation (acceptability of decisions taken) (Weber 2016b, 251). These basic values should be the foundation for appropriate legitimacy strategies, but the schemes must be broad enough and leave room for adaptations in a given context.

Forms and Legal Framework of Decision-Making

The concept of multistakeholderism requires at least two classes of stakeholders (Raymond and DeNardis 2015, 572, 575). Different concepts of multistakeholderism can be and are implemented in reality, subject to the types of actors that are involved and the nature of authority relations between these actors.

Depending on the design of the actors and the scope of relations, the combinations in a matrix can be numerous (Raymond and DeNardis 2015, 577, 583). Furthermore, multistakeholder arrangements usually also vary by level. Four ideal-typical structural models have been developed: hierarchy (for example, the case of the International Telecommunication Union), homogeneous polyarchy (for example, the Internet Engineering Task Force, W3C, and International Organization of Securities Commissions), heterogeneous polyarchy (for example, ICANN, the UN Global Compact), and anarchy (ibid., 580, 603). Often, the choice of the models is limited, but some discretion for the involved stakeholders is mostly given.

In general, a multistakeholder decision-making framework should encompass the following main elements:

- Identification of the most adequate set of stakeholders participating in a particular issue
- Definition of the criteria and mechanisms for the selection of representatives from different groups
- Avoidance of capture of multistakeholder processes by corporate power or influential nongovernmental organization
- Implementation of crowdsourcing techniques bringing inputs into dialogue on difficult topics
- Establishment of technologies helping the representatives liaise with their constituencies and monitor reached agreements
- Creation of a technological framework facilitating dialogue to reach a minimum consensus in a multistakeholder body
- Methods for accelerating the decision-making processes in multistakeholder bodies
- Theoretical models supporting consensus building and decision-making in multistakeholder environments (Almeida, Getschko, and Afonso 2015, 74, 78)

In designing the multistakeholder decision-making framework, political context and cultural factors must be taken into account (Weber 2016b, 250). The implementation should also consider the effect of existing standards on the decision-making of an organization and whether to lower potential entry barriers for stakeholders (see also Van Huijstee 2012, 45).

Concretization for Internet Governance

Practical experiences have shown over the last few years that a range of approaches, mechanisms, and tools are available for the realization of multistakeholder objectives, leading to the acknowledgment that a toolbox should be developed with a number of suitable instruments (Gasser, Budish, and West 2015, 2; see also Buzatu 2015, 11–14). This assessment is not surprising since multistakeholder models must rely on an ever-increasing participation by those with interests, capacities, and needs (Doria 2013, 115, 135). Therefore, the multistakeholder concept may not be seen as a value in itself to be applied homogeneously to governance functions—that is, it is not a one-size-fits-all solution (Weber 2016b, 258). However, the development of systems for sharing information, taking decisions, designing checks and balances, and implementing assurance models is at the heart of effective multistakeholder initiatives (Buzatu 2015, 16).

Multistakeholderism is practiced in reality in, for example, the context of the Internet Governance Forum, which includes a special committee, the Multistakeholder Advisory Group (MAG), whose roughly 40 seats represent the five world regions and also balance gender (Hofmann 2016, 16). The multistakeholder element, addressing participation in different ways and using different terms, also prominently appears in the NETmundial Multistakeholder Statement released at the closure of the NETmundial Conference held in São Paulo in April 2014.[3] Attendees from around the world, in government, the private sector, civil society, the technical community, and academia, crafted this nonbinding statement. In the meantime, ICANN also partly opened the door for some multistakeholder exchanges, mainly in connection with accountability.[4]

Without any doubt, the debates about Internet governance and multistakeholderism must encompass the general and relevant policy issues, in particular legitimacy, transparency, and accountability; topics have been linked only in a limited way (Weber 2016b, 259–262; see also Gasser, Budish, and West 2015, 10–11, 22–23, 26). In addition, topics such as decision-making procedures (Zingales and Radu 2017, 53, 67), formation and operation inclusiveness, and effectiveness need further attention (Weber 2016b, 262–264; see also Gasser, Budish, and West 2015, 11–13, 18–26).

In view of these manifold factors, no standard way to form multistakeholder groups can be established. Depending on the cultural and the contractual factors in shaping the functioning and the outcome of governance groups (for example, the preexisting relationships between the stakeholders, the relationship between the governance group and the governmental institution, the allocation of resources, and geopolitical factors), the dimensions of multistakeholder groups must be designed; therefore, a broad spectrum of purposes can be listed, ranging from open-ended missions to issue-specific tasks (Gasser, Budish, and West 2015, 10, 25; Weber 2016b, 258).

Even if multistakeholderism is not a value as such, it must be considered as a possible approach for meeting salient public interest objectives by determining what types of decision-making are optimal in the given functional and political context (Raymond and DeNardis 2015, 610). The following elements and action points support effective multistakeholder governance (Buzatu 2015, 28–31; Weber 2016b, 265):

- Identification and articulation of purpose and objectives (appropriate setting of the stage)
- Identification of the players (adequate and precise definition of the stakeholders)
- Development of the applicable multistakeholder governance model
- Definition of the envisaged procedural formation and operation principles and description of the scope of inclusiveness
- Determination of the appropriate level of transparency
- Implementation of accountability standards
- Provision of guidance for the implementation of the agreed standards
- Identification of a sustainable and credible funding model for the multistakeholder processes
- Development of oversight and assurance mechanisms

In a nutshell, multistakeholder initiatives can be seen as fora multipliers through manifold platforms for dialogue. Furthermore, such initiatives are suitable to establish fora for evolving standards and governing mechanisms (on the human rights issue, refer to Braman's chapter 2 and Jørgensen's chapter 8). But many factors in multistakeholder initiatives need further research; in particular a multidisciplinary examination of the relevant questions incorporating socio-legal, economic, policy-oriented, and game theory studies, as well as interdisciplinary information studies drawing on political analyses appears to be indispensable (Brown and Marsden 2013, 200–201). Developing a multidisciplinary catalog of methodologies as well as the corresponding multidisciplinary tools can improve the chances for the existence of an appropriate tool kit as well as the comprehension of challenges of better participative decision-making and configuration of governance concepts (Weber 2016b, 265).

Conclusion

If a legal lens is applied to Internet governance, then the original quite soft law has become much harder. Normative principles gain importance, and quality criteria of regulation are more closely taken into account.

On the one hand, the legal framework has been strengthened concerning the substantive Internet governance principles such as legitimacy, transparency, and accountability. The described developments in the legislative context show the awareness of the involved stakeholders in broadening the foundation for policy decisions. In addition, the more frequently chosen multistakeholder approach has been concretized, and its applicable elements are hammered out to achieve an appropriate regulatory environment.

On the other hand, sovereignty considerations have gained much higher attention in many countries—that is, national control of infrastructure becomes increasingly important. As a consequence, legal fragmentation causes obstacles that jeopardize cross-border data flows (see Braman's chapter 2; Mueller 2017; Voelsen 2019). The politically envisaged fragmentation contradicting the objective of legal interoperability (see Palfrey and Gasser 2012; Weber 2014a) must mainly be seen as a power struggle over the future of national sovereignty in the digital world (Mueller 2017, 5).

These new movements increase the challenges for the legal profession. More interdisciplinary research is imperative, thereby widening the perspective for an overarching perception of Internet governance in all social sciences (for a good example, see Kerr, Musiani, and Pohle 2019). Following Rousseau's social contract concept, a legal framework for Internet governance must establish open communication channels leading to "commons of knowledge" (Weber 2016a, 214). As a consequence, policy makers are now confronted with cross-sectoral concerns simultaneously combining socioeconomic, political, and ethical dilemmas. The capacities of the current institutional patchwork to deliver on transborder challenges have become questionable, not at least because other issues such as climate change or international migration flows determine the political agendas (Radu 2019, 196). Therefore, advocates of Internet governance are called on to regain confidence in their willingness to uphold a normative framework that guarantees that rules for Internet use are designed in a way to be trustworthy in the eyes of civil society.

Notes

1. The most influential and ambitious attempt to describe an interchange paradigm goes back to Parsons (1991), who distinguishes four social subsystems: adaptation, goal attainment, integration, and pattern-maintenance/latent tension management (AGIL).

2. The relevance depends on the circumstances: fundamental principles (for example, human rights) are less likely to be subject to substantive adaptation.

3. The statement is available at http://netmundial.br/netmundial-multistakeholder -statement/.

4. For an overview of the information given by ICANN on accountability, see https://www.icann.org/resources.accountability.

References

Almeida, V., Getschko, D., & Afonso, C. (2015). The origin and evolution of multi-stakeholder models. *IEEE Internet Computing, 19*(1), 74–79.

Amstutz, M. (2011). Mechanisms of evolution for a law of the future. In S. Muller, S. Zouridis, M. Frishman, & L. Kistemaker (Eds.), *The law of the future and the future of law* (pp. 395–405). Oslo, Norway: TOAEP.

Baldwin, R., Cave, M., & Lodge, M. (2012). *Understanding regulation: Theory, strategy, and practice*, 2nd ed. Oxford, UK: Oxford University Press on Demand.

Barlow, J. P. (1996). *A declaration of the independence of cyberspace*. Retrieved from https://www.eff.org/cyberspace-independence

Berman, P. S. (2007). Global legal pluralism. *Southern California Law Review, 80*; Princeton Law and Public Affairs Working Paper No. 08–001. Retrieved from https://ssrn.com/abstract=985340

Brown, I., & Marsden, C. T. (2013). *Regulating code: Good governance and better regulation in the information age*. Cambridge, MA: MIT Press.

Buzatu, A.-M. (2015). *Multi-stakeholder approaches to governance: Challenges and opportunities*. DCAR Horizon Working Paper 8.

Chik, W. B. (2010). "Customary Internet-ional law": Creating a body of customary law for cyberspace. Part 1: Developing rules for transitioning custom into law. *Computer Law & Security Review, 26*(1), 3–22.

Clark, I. (2005). *Legitimacy in international society*. Oxford, UK: Oxford University Press.

de Búrca, G., Keohane, R. O., & Sabel, C. (2014). Global experimentalist governance. *British Journal of Political Science, 44*(3), 477–486.

DeNardis, L. (2014). *The global war for Internet governance*. New Haven, CT: Yale University Press.

Doria, A. (2013). Use [and abuse] of multistakeholderism in the Internet. In R. Radu, J.-M. Chenou, & R. H. Weber (Eds.), *The evolution of global Internet governance* (pp. 115–138). Zürich, Switzerland: Schulthess.

Gasser, U., Budish, R., & West, S. M. (2015). *Multistakeholder as governance groups: Observations from case studies* (Series 1). Berkman Center for Internet & Society Research Publications. Retrieved from https://papers.ssrn.com/sol3/papers.cfm?abstract_id=2549270

Habermas, J. (1992). *Faktizität und Geltung*. Frankfurt, Germany: Suhrkamp.

Hofmann, J. (2016). Multi-stakeholderism in Internet governance: Putting a fiction into practice. *Journal of Cyber Policy, 1*(1): 29–49.

ICANN/WEF Panel on Global Internet Cooperation and Governance Mechanisms. (2014). *Panel Report: Towards a Collaborative, Decentralized Internet Governance Ecosystem.* Retrieved from https://www.icann.org/en/system/files/files/collaborative-decentralized -ig-ecosystem-21may14-en.pdf

Jørgensen, R. F. (2013). *Framing the net: The Internet and human rights*. Cheltenham, UK: Edward Elgar.

Kerr, A., Musiani. F., & Pohle, J. (2019). Communication and Internet policy: A critical rights-based history and future. *Internet Policy Review, 8*(1), 1–16.

Klimburg, A. (2013). *The Internet Yalta*. Center for a New American Security. Retrieved from https://s3.amazonaws.com/files.cnas.org/documents/CNAS_WCIT_ commentary-corrected-03.27.13.pdf

Luhmann, N. (1975). *Legitimation durch Verfahren*, 2nd ed. Darmstadt/Neuwied, Germany: Suhrkamp.

Mahler, T. (2014). A gTLD right? Conceptual challenges in the expanding Internet domain namespace. *International Journal of Law and Information Technology, 22*(1), 27–48.

Mahler, T. (2019). *Generic top-level domains: A study of transnational private regulation*. Cheltenham, UK: Edward Elgar.

Mueller, M. (2017). *Will the Internet fragment? Sovereignty, globalization and cyberspace*. Cambridge, UK: Polity Press.

Murray, A. (2007). *The regulation of cyberspace: Control in the online environment*. Abingdon, UK: Routledge-Cavendish.

Organization for Economic Cooperation and Development. (2005). *Guiding principles for regulatory quality and performance*. Paris, France: Author.

Ost, F., & van de Kerchove, M. (2002). *De la pyramide au réseau?: pour une théorie dialectique du droit*. Brussels, Belgium: Presses de l'Université Saint-Louis.

Palfrey, J., & Gasser, U. (2012). *Interop: The promise and perils of highly interconnected systems*. New York, NY: Basic Books.

Parsons, T. (1991). *The social system*. New York, NY: Taylor and Francis.

Post, R. (1991). Introduction: The relatively autonomous discourse of law. In R. Post (Ed.), *Law and the order of culture* (pp. vii–xvi). Berkeley: University of California Press.

Radu, R. (2019). *Negotiating Internet governance*. Oxford, UK: Oxford University Press.

Raymond, M., & DeNardis, L. (2015). Multistakeholderism: Anatomy of an inchoate global institution. *International Theory, 7*(3), 572–616.

Reed, C. (2012). *Making laws for cyberspace.* Oxford, UK: Oxford University Press.

Rioux, M. (2014). Competing institutional trajectories for global regulation: Internet in a fragmented world. In R. Radu, J.-M. Chenou, & R. H. Weber (Eds.). *The evolution of global Internet governance* (pp. 37–55). Zürich, Switzerland: Schulthess.

Senn, M. (2011). *Non-state regulatory regimes: Understanding institutional transformation.* Heidelberg, Germany: Springer Science & Business Media.

van Huijstee, M. (2012). Multi-stakeholder initiatives: A strategic guide for civil society organizations. Amsterdam, Netherlands, Working Paper. Retrieved from http://papers.ssrn.com/sol3/papers.cfm?abstract_id=2117933

Voelsen, D. (2019). Risse im Fundament des Internets: die Zukunft der Netz-Infrastruktur und die globale Internet Governance. Berlin, Germany, Working Paper. Retrieved from https://www.ssoar.info/ssoar/handle/document/62982

Weber, R. H. (2002). *Regulatory models for the online world.* Zürich, Switzerland: Schulthess.

Weber, R. H. (2009). *Shaping Internet governance: Regulatory challenges.* Zürich, Switzerland: Schulthess.

Weber, R. H. (2013). Visions of political power: Treaty making and multistakeholder understanding. In R. Radu, J.-M. Chenou, & R. H. Weber (Eds.), *The evolution of global Internet governance* (pp. 95–113). Zürich, Switzerland: Schulthess.

Weber, R. H. (2014a, December). Legal interoperability as a tool for combatting fragmentation (Paper Series No. 4). Waterloo, Canada: Global Commission for Internet Governance.

Weber, R. H. (2014b). *Realizing a new global cyberspace framework.* Zürich, Switzerland: Schulthess.

Weber, R. H. (2016a). Elements of a legal framework for cyberspace. *Swiss Review of International and European Law, 26*(2), 195–215.

Weber, R. H. (2016b). Legal foundations of multistakeholder decision-making. *Zeitschrift für Schweizerisches Recht, 135*, 247–263.

Weitzenboeck, E. M. (2014). Hybrid net: The regulatory framework of ICANN and the DNS. *International Journal of Law and Information Technology, 22*(1), 49–73.

Working Group on Internet Governance. (2005). Report of the working group on Internet governance. Retrieved from http://www.wgig.org/docs/WGIGREPORT.pdf

Zingales, N., & Radu, R. (2017). In search for the holy grail: Meaningful multistakeholder governance in Internet policy-making. In C. Prins, C. Cuijpers, P. L. Lindseth, & M. Rosina (Eds.), *Digital democracy in a globalized world.* Cheltenham, UK: Edward Elgar.

6 Web Observatories: Gathering Data for Internet Governance

Wendy Hall, Aastha Madaan, and Kieron O'Hara

The World Wide Web is the most significant application of the Internet, a simple, easy-to-use information space indexed by uniform resource identifiers on which are built most of the services accessed by Internet users today, including search engines, video streaming, and social media. Although the Internet predated the web by decades, the web brought a technical revolution, with the most significant social impact changing many aspects of how communication takes place and social behavior is shaped in politics, economics, leisure, entertainment, scientific research, commerce, and social interaction. As noted by Mueller and Badiei in chapter 2, the creation of the web (together with the invention of the browser) was partly responsible for the emergence of the Internet as a mass public medium, which has made Internet governance such a key issue. It has evolved from an efficient, although not unique, document repository to an active socio-cognitive space in which people express ideas and emotions across geopolitical boundaries. Its impact and reach have emphasized the importance of interpersonal integrity and issues of data ownership, privacy, trust, and surveillance in mainstream research, while the data, tools, and platforms that constitute and enable the web are distributed geographically, across legal jurisdictions. They are also used differently, with different levels of effectiveness and embeddedness, by diverse cultures, genders, age cohorts, and economic classes.

The social challenges of context of use and analysis make the task of studying the web complex. Neither are the technical challenges trivial, as the data that are generated may or may not be preserved. Rapid changes make it harder for researchers to find evidence to test hypotheses related to research questions pertinent to governance and policy making. For example, consider the prominent roles played by private social networking entities in political campaigns run on the web (Stieglitz and Dang-Xuan 2013) or during disasters (Ngamassi, Ramakrishnan, and Rahman 2016) when these platforms play a critical role in providing physical aid and basic living facilities to the stakeholders in consideration. In one specific example, Facebook has been implicated in the apparent violence directed at the stateless Rohingya people in Myanmar (2016–2018), as a means of orchestrating anti-Rohingya sentiment (Banyan 2017), as a means for jihadis to spread extremist messages (Singh and Haziq 2016; Stevens and O'Hara 2015), and as a means for the Rohingya to inform the world of their problems (Wong, Safi, and Rahman 2017). The effects of social networking may be found in more stable societies too—for instance, the rise of smartphone use among US teenagers correlates with (though may or may not have caused) a sudden rise in suicide rates over the same period by 31 percent for teenage boys and more than 100 percent for girls (Twenge et al. 2018).

In this chapter we argue that, while it is important to address various Internet governance issues at the protocol level (or design level), it is also critical to understand the affordances of Internet use for the interactions among stakeholders. For that, effective, ethical, and secure methods of gathering and sharing data will be required. In this chapter, we consider the challenges to creating and disseminating such methods and describe an architecture for that purpose, which we call the Web Observatory (Hall et al. 2013; Tiropanis et al. 2013). While the Web Observatory is situated in the web context, using web protocols to organize data, its value for Internet governance stems from the importance of the web as the gateway to the use of the Internet. The architecture is designed to meet a set of technical, social, and legal challenges that will stand in the way of any kind of evidence-based Internet governance. Additionally, although we don't highlight this in this chapter, the idea of a Web Observatory is intended also to provide a pragmatic means to facilitate the sharing of data; to this end, it has been argued that the Web Observatory is a potential architecture for a *data trust* (O'Hara 2019), identified as a key enabler of the growth of the artificial intelligence industry (Hall and Pesenti 2017).

Governance, Content, and Data

The web is a socio-technical construct (Hall and Pesenti 2017), and as such its effects ripple through its embedding societies as emergent macro-level phenomena such as the formation of "crowds" as a response to real-world events, the spread of emotional contagion, the site of opinion markets that affect the results of a real-world event such as an election, and the host of data marketplaces where data can be traded for monetary benefits while compromising personal privacy. It follows that, on top of the web's technical development and its open standards, those with an interest in governance need also to consider content usage and access patterns to understand issues highlighted by DeNardis (2010)—digital equality (Hargittai 2010), social media (boyd 2008; DeNardis and Hackl 2015), identity (Turkle 2005), knowledge production (Benkler 2006), Web 2.0 critiques (Lanier 2010), and copyright and information intermediaries (Vaidhyanathan 2007). Data can provide deeper insights about large-scale populations in real time, opening out research questions about the social behavior germane to a user-centric view of Internet governance (Dutton and Peltu 2005), on issues such as social and political movements; political participation and trust; crisis prevention, preparedness, response management, and recovery; individual, group or community, and national identities; and personal, local, national, and global security (De Roure 2014). To take one example, Wikipedia's underlying technical platform remains more or less the same as it was in the beginning, but the way people interact through it has varied significantly over the years (Kittur et al. 2007), and so the regulation of Wikipedia cannot simply be a matter of developing protocols.

A rich literature on Internet governance is available that describes various perspectives, including governance of platforms, layers of Internet infrastructure and applications where governance needs to be separated, distributed governance based on geopolitical boundaries, and data governance. Some of these also highlight how effective governance policies require working at a variety of scales, from the micro level of detail of individual protocols like HTTP (hypertext transfer protocol)[1] or HTML (hypertext markup language)[2] to macro-level emergent behavior such as blogging, spamming, or e-commerce and how the social phenomena emerge onto, diverge within, and submerge into it.

For example, Google knows much more about us than we know about Google (Hall and Pesenti 2017). Even if we are not involved in any power play or asymmetrical encounter with Google, such asymmetry has ramifications. This has sparked debate about user privacy, unwanted profiling of customers (people) by technical giants, and tracking of their online activities, preferences, and location (Hildebrandt 2016; Zuboff 2019). These mandate governance policies to be inclusive of these new forms of interactions and relationships on the web.

Because of these social and political imperatives, a shift in the traditional Internet governance view has been proposed in the Tunis Agenda, successor the UN Working Group on Internet Governance, and elsewhere (Wagner 2016). Whereas the traditional view of governance focused more on the technical functions and standards required to keep the Internet open, unfragmented, stable, resilient, and secure, the UN group advocates "the development and application by governments, the private sector and civil society, in their respective roles, of shared principles, norms, rules, decision-making procedures, and programmes that shape the evolution and use of the Internet" (Wagner 2016). This will become more critical because of the rapid change and maturing of information technologies, including machine learning, natural language processing, face recognition, robotics, blockchains, and cryptocurrencies. Ultimately these technologies will converge with the Internet of Things (IoT), and as they do the interrelationship between humans and machines will pose unprecedented challenges for human societies and how they are governed (Fry et al. 2015). Meanwhile, many social activities increasingly take place online using the functionality of networks via commonly used connected devices, creating what have been called social machines, which themselves generate important quantities of data (Shadbolt et al. 2019).

Hence, an infrastructure is critical to overcoming the main barrier of web data collection and analytics essential for evidence-based study on the web. In addition, the speed at which data interactions occur on the web means that the data become obsolete and outdated at a rapid pace even if regular snapshots are taken (Hall et al. 2013).

Governance Challenges for a Distributed Data Infrastructure

The Need for an Infrastructure

Data have become the new fuel empowering decision-making in almost every sector. Governments make a wealth of public data openly available (on sites such as www.data.gov, United States; www.data.gov.uk, United Kingdom; data.gov.in, India; and www.data.gov.au, Australia) under different licensing based on their use. But observing a restricted number of siloed datasets that are deemed nonsensitive provides a narrow view to any problem or issue. The real value of data comes from the broader perspective of multiple datasets brought together around a specific question or issue and from a range of sources (Fry et al. 2015), which suggests the idea of a platform to bring together data for the purpose of analysis and interrogation that furnishes methods and tools that enable researchers to locate, analyze, compare, and interpret information consistently and reliably (Hall et al. 2013; Hall et al. 2014; Tinati et al. 2015; Tiropanis et al. 2013).

It goes without saying that a centralized data store cannot possibly scale; the model proposed here is a platform where anyone with a dataset who wishes (and has the right) to share it could display the metadata to enable the data to be discovered by potential users. We call such rights holders data owners as a shorthand—they include data controllers whose data are personal under European data protection legislation, although the data shared via the Web Observatory need not be personal. Such data owners could, in principle, be anyone able to exercise rights to share, including governments, corporations, nongovernmental organizations, educational institutes, scientists and academics, or even private individuals (for instance, someone who wished to consider sharing his or her medical data with selected health care providers or researchers). However, joining the Web Observatory and advertising metadata does not mean that the data owner is obliged to share; the owner remains in control (and in particular, a controller of personal data would remain the data controller) and makes the decision to share. The Observatory is a distributed infrastructure to enable data discovery and sharing, not an automatic distributor of data. Hence, sharing of the data takes place *only* when the dataset owners receive a request and *only* in accordance with the owners' constraints, legal and ethical requirements, and business models. The governance of individual datasets remains with their owners; the responsibility of the infrastructure would be to provide protocols to support data discovery and sharing. To answer open questions and ensure open access, institutions owning data, or with common interests in sharing data, need to come together within the Web Observatory to build an active community engaging in experimentation and innovative problem-solving, involving the generation and sharing of both qualitative and quantitative data for evidence-based decision-making (Verhulst et al. 2014).

When data are especially sensitive (and this is not only personal data— consider, for instance, geologic data relevant to a fracking inquiry, which raises no privacy concerns), it may be that the data owner does not consider sharing it at all. The owner might accept and process queries from third parties and return the results, perhaps using differential privacy techniques to ensure that the queries in aggregate aren't disclosive. Or the owner may even refuse access to the data but may post visualizations of results from processing.

Indeed, any data owner could allow access to visualizations of data whether sensitive or nonsensitive or access to tools or techniques that have proved valuable for data analysis or visualization. The Observatory can be the repository for anything valuable to the data analytics community, not simply metadata about valuable datasets. So, for example, the metadata describing transcripts of sessions of the Internet Governance Forum described by Cogburn in chapter 9 might be posted to the Web Observatory; those who could add value to those data could contact, via the Observatory, the rights holder and ask for access. Similarly, the mailing lists whose analysis is described by Ten Oever, Milano, and Beraldo in chapter 10 could be discovered. The mailing lists may be sensitive, so sharing might take place only under very specific conditions (e.g., the terms and conditions of access might preclude publishing direct quotes). Meanwhile, limited access could be granted to the commercially sensitive data held by major search engines and social media companies, discussed by Jørgensen in chapter 8—if not to the data itself, at least to summaries, statistics, and visualizations, or perhaps such companies might accept a limited number of queries from academics and policy makers without compromising any comparative advantage the data are perceived to confer. This could be especially valuable for Internet governance, given the importance that such sites have in the ecosystem of the Internet. The value of this kind of eclecticism would help address difficulties in studying the private sector, multistakeholder governance, and overstudying open systems (see DeNardis's chapter 1).

Though a distributed data infrastructure could and should be open to private and public actors, an early win could be the augmentation of current government digital platforms and open datasets. Increasingly, digital governments require access to data services that are internally and externally produced, which often creates a complex ecosystem of social systems (Tinati et al. 2015). Thousands of datasets are now openly available through open data portals describing local businesses, city-sensor networks, and live transport and traffic data. But not all government data can be opened for access so easily, and not all the data governments use are generated or curated by them, what with outsourcing and other policy delivery partnerships with government. In this context, there is an opportunity for a wrapper to expose and share government data, which importantly allows third parties to access and produce analytical and visual representations of the data while still retaining access control (Tinati et al. 2015). Furthermore, a data infrastructure would offer an opportunity for governments to link to nongovernmental datasets, enriching existing data and providing new insights not originally envisioned (Hall et al. 2013; Tiropanis et al. 2013).

From the point of view of Internet governance, governments and nongovernmental actors could observe socio-technical phenomena from such an infrastructure via data from domains using web protocols (e.g., social media, crowdsourcing platforms, health care, city sensors). Data could be made available for those needing to make evidence-based policy decisions relating to Internet governance without exposing it as open data or indeed without the data owner having to surrender control over access at all. The Observatory provides the means for discovery and communication between potential consumers of the data and their owners and may itself be open or closed to new members. To the extent that the data are sensitive, the Web Observatory will inherit all the legal, ethical, and political issues relating to sharing sensitive data. The Observatory infrastructure does not solve these problems, but because control over the data remains with the data owners, neither does it create new information flows over which nobody has any control. In particular, it remains the decision of the data owners whether to share sensitive data (e.g., about a smart city) with someone who has requested access, to restrict its use for Internet governance purposes only, or alternatively, to share with a wider range of data consumers. To the extent that potentially sharing data about socio-technical phenomena has implications for wider society, the proposed Observatory infrastructure minimizes those implications. As a consequence, data governance, access to datasets, usage tracking to understand derivations of datasets, and user trust are major concerns for such an ecosystem. In the rest of this section, we describe these concerns in detail.

Challenges

An important aim is to provide strong support for *user-centric transparency* by enhancing end users' awareness about their choices for data sharing, dependent on interactions with other stakeholders and the sensitivity of datasets, requiring case-by-case analysis of the use of data. Crafting the terms and conditions of data sharing requires addressing the questions of liability and responsibility of data sharing on the part of operators, data-sharing companies, and other stakeholders. The growing imbalance of power created between citizens and companies through the privileged access that corporations have to information on our collective social lives is set to become an increasingly pressing social and political issue (Davies 2013; O'Hara 2015; and see also Jørgensen, chapter 8). Some key challenges need to be addressed by anyone proposing to share data:

Privacy is critical for understanding data processing on shared datasets of personal data (or data derived from personal data), with cultural context an important consideration. Global rules may not be applicable, or sufficient, at the local scale (O'Hara 2019).

Open data and the field of data integration generally has raised questions of interoperability, transparency, accountability, and reusability.

Trust in the integrity of data is becoming even more important with auto-mated data collection through smart devices and analytics being performed by a range of actors (even before we consider issues concerning the increasing tendency of malign actors to attempt to mislead—an economy that supports fake news farms could equally support fake data farms). Understanding the provenance of data and measuring its quality are key challenges.

Sovereignty has made the questions of where data are stored and accessed especially relevant in the context of cloud storage, especially in light of increasing moves toward data nationalism (Chander and Lê 2014–2015).

Uncertainty about issues such as liability if data are misused and the secu-rity of data following a share has been identified as a key blocker to data sharing (Hall and Pesenti 2017).

An infrastructure for amassing data for evidence-based reasoning about Internet governance will need to bring together diverse communities for data sharing and reuse across geopolitical and application contexts and across public, private, and nonprofit sectors. These are questions of cor-porate social responsibility, as well as the ethical and legal consequences potentially faced by corporations involved in these controversies (Dutton and Peltu 2005). The challenges laid out here are even more critical with the rapid growth of technologies such as the IoT and artificial intelligence, which promote machine-to-machine interactions and capture a huge amount of potentially personal data about individuals.

Data that are personal lead to much concern over the expansive digital surveillance practices of some countries, the rise of nation-specific data-localization policies that cite privacy concerns as a justification for requiring providers to store data within national borders (Chander and Lê 2014–2015), and emerging and controversial policies such as the right to be forgotten ruling in the European Union (DeNardis and Hackl 2015; O'Hara and Shadbolt 2015; O'Hara, Shadbolt, and Hall 2016). The emphasis of the EU General Data Protection Regulation on a notion of data protection delineated by a rich jurisprudence contrasts with countries like India that have yet to define privacy and countries like China where paternalism plays a larger part in policy than supporting the rights of autonomous individuals. In a data infrastructure, individual units are likely to align to local policies and laws, which may reduce legal concerns but at the same time raises the concern of interoperability across the network (O'Hara and Hall 2018).

Data, communications metadata, or data about networks provide a lot of actionable and personally identifiable information, especially when augmented with auxiliary data. The routinization of extensive metadata collection as well as contextual content analysis is a fundamental departure from the Internet's original end-to-end design (of locating intelligence at end points, technical neutrality toward packet contents, and using IP addresses simply as virtual identifiers) (DeNardis 2014). Moreover, identification technologies also collect information on the hardware for using the platform (device information) and software information, including browser type (software footprint) (DeNardis 2014). To address concerns about this, which will be essential to foster wider trust of the Web Observatory infrastructure, each resource on the infrastructure would need to be annotated with appropriate metadata to restrict access and use of the dataset and have the ability to notify the data owners and publishers in real time about possible privacy breaches and would need to be protected by a clear and understandable set of terms and conditions to prevent disclosure of personal information.

Finally, we note that if people are to have any control over their personal data, they need rights over the data and transparency about what is happening to it. But the exercise of these individual rights is truly effective only if an organization's technology is fully responsive to them and has the right functionality embedded in it. The core individual data protection rights in the General Data Protection Regulation are the "right of access," "right to rectification," "right to erasure" (or the right to be forgotten), "right to restrict processing," "right to data portability," and "right to object" (GDPR Individual Rights [n.d.]). In a functional sense, these rights require the technology to connect individuals to their personal data, the technology to classify the data with respect to the purpose of use or processing and for mapping the full data life cycle, the technology to make it searchable, and the technology to rectify it and perform erasure and anonymization. These are complex requirements and may not easily be incorporated, but at a minimum one would expect the implementation of provenance-tracking mechanisms to track data and application use to bring some measure of accountability to the system.

The Web Observatory

Clearly, the development of such a global infrastructure for data sharing is a major research question. In this section, we describe the concept of a web observatory that meets the infrastructure requirements set out in the previous section, as a distributed global resource in which datasets, analytical tools, and cross-disciplinary methodologies can be shared and combined to foster web science: the emerging interdisciplinary field of the study and engineering of large-scale distributed information systems, particularly the World Wide Web, with a focus on their cocreation by human users and participants (Hall et al. 2014; Shneiderman 2007; Tiropanis et al. 2013). The aim of the Web Observatory is to provide a locus for subject-centric management of data, to complement the current paradigm of organization-centric data management (Van Kleek et al. 2014) and support longitudinal web-science analysis.

Hall et al. (2013) proposed the idea of the Web Observatory[3] to bootstrap the analytics to create evidence of newer concerns emerging from collated and archived data on the web. It provides the capability to curate datasets and aggregate interaction data (web clickstreams, dialogue from crowd-sourcing platforms), descriptive data (demographics and geospatial data), behavioral data (usage history), and attitudinal data (opinions, preferences of stakeholders) (De Roure 2014; Hall et al. 2013; Tiropanis et al. 2013). It provides infrastructural support in the form of analytical tools and datasets to investigate methods and mechanisms by which people, as a collective society, could be effectively studied in academic research settings through analysis of the archival information traces they created online (Van Kleek et al. 2014). It enables users to share data with each other, while retaining control over who can view, access, query, and download their data (Tinati et al. 2015). Web Observatory infrastructure would certainly enable the evolution of social machines as means of bottom-up coordination of problem-solving at scale (in terms of both providing the data to enable the academic study of social machines and supporting their operation where data are needed as input [Shadbolt et al. 2019]).

A web observatory infrastructure can be a generic set of desiderata. However, in this section we report on an attempt to implement these ideas in a specific system, *the* Web Observatory, with a definite article. The effort to create the Web Observatory has involved many different communities and organizations, including the Web Science Trust network of laboratories (WSTNet),[4] other major research groups in this area, government agencies, public sector institutions, and industry (Hall et al. 2016; Price et al. 2017). The Web Observatory has formed partnerships with the World Wide Web Consortium (W3C), the Open Data Institute (ODI) in the United Kingdom, the Fraunhofer Institute for Open Communication Systems (FOKUS), the Web Foundation, and a growing list of industrial collaborators (Tiropanis et al. 2013). Also, researchers affiliated with the Web Observatory at the University of Southampton have cooperated with the University of South Australia to support data-driven government policy making for the Adelaide government (Fry et al. 2015). The Web Observatory, which is based at Southampton, has provided access to multiple government and academic datasets to answer questions concerning the provision of public services on the basis of the demographic landscape of Adelaide's neighborhoods.

The Web Observatory contains metadata, catalogs of data, visualizations, analyses, and tools. Hence, if standards were established to express metadata (e.g., using linked data formalisms), the Web Observatory could be distributed in a network, each node of which could be located remotely, because of the interoperability that would result. The Web Observatory therefore operates as a decentralized, distributed infrastructure for sharing data and analysis (Tinati et al. 2015; Tiropanis et al. 2013) on the Web. It is, in other words, a network of web observatories (we refer to observatories within the network as *nodes*). The Southampton University Web Observatory (SUWO) mentioned in the previous paragraph is one node of the network.

Appropriate standards to enable the discovery, use, combination, and persistence of datasets and tools are being developed in the W3C Web

Observatory community group[5] (Hall et al. 2014), established to foster discussions on standardization requirements essential to enable interoperability between available resources and on identifying the opportunities for industry and global government agencies to contribute large-scale systems, expertise, and datasets.

Architecture of the Web Observatory

Figure 6.1 describes the different architectural components of the Web Observatory. The Web Observatory harmonizes data sources from, for example, social media, open data repositories, and Internet archives. An interoperable catalog makes the metadata available in compliant standards. These data are then available to researchers through the visualization and analytical tools and methodologies available on Web Observatory nodes. In making this web of observatories possible, the following principles are required in the deployment of Web Observatory nodes or instances (Tiropanis et al. 2014):

- Resources related to the Web Observatory (projects, datasets, analytical applications, and people) need to have unique identifiers, preferably uniform resource identifiers or application programming interface (API) access points.

- There need to be explicit links between analytical applications and the datasets that they use.

- There need to be explicit links between Web Observatory resources and related use, scholarship, and discourse.

- Metadata should be published for all available resources in a Web Observatory instance to enable search.

- Datasets and analytical applications hosted or listed on a Web Observatory node can be public or private; the publisher needs to control who gains access to them.

- It should be possible to enable access for identified individuals (or to applications using their credentials) to specific datasets or applications hosted in local or remote datasets.

- It should be possible to support distributed queries across Web Observatory instances and to make computational resources on each instance available to that end.

The Web Observatory architecture is implemented as a reverse proxy server[6] enabling the data owners to host datasets at their local sites but also

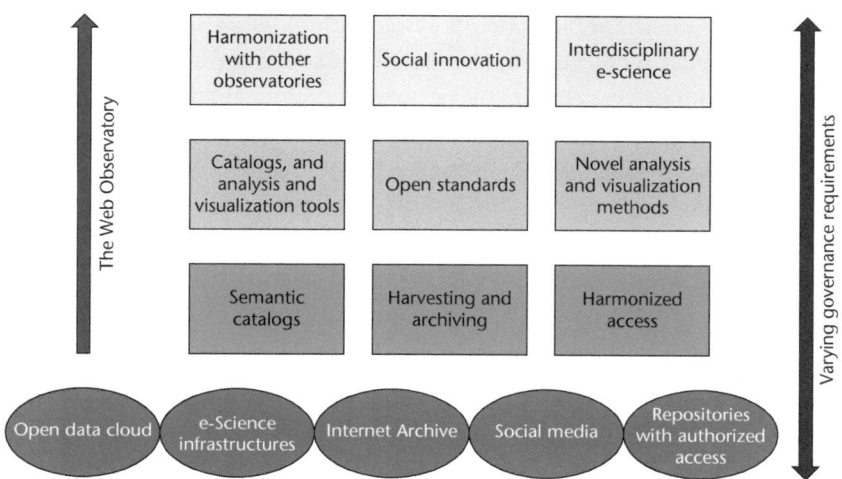

Figure 6.1
Web Observatory components and governance requirements.

allowing the observatory users to discover it through metadata search (Price et al. 2017; Wang et al. 2017). It adopts a decentralized architecture in which resources can be accessed in situ without the additional need of gathering them centrally. Users can publish to the Web Observatory resources from outside, and data generated within the Web Observatory such as system logs and user information are made available as datasets and listed in the Web Observatory itself (Siow, Wang, and Tiropanis 2016). The discoverability of the metadata is defined by the data owner or publisher on the Observatory as private or public. A Web Observatory node may be initiated by individuals for use as a personal observatory (Siow, Tiropanis, and Hall 2016). Different observatory nodes are not obliged to host the same software but should comply with the underlying principles of the Web Observatory. A distributed search tool is available to allow the end users to search for datasets and tools across the distributed Web Observatory using the metadata attributes.

The following core functionalities of the Web Observatory infrastructure support engagement across communities and emphasize ethical and legal rectitude.

Metadata of datasets and analytical tools   Metadata describing the listed resources and projects is published. In this way, descriptions of resources can be harvested and listed in other Web Observatory nodes or web-based resources. If a listed application (or visualization) uses a listed dataset, the link is explicitly mentioned in the metadata for the application. Metadata not only facilitate accurate discovery but also provide links between related resources. For example, datasets and the analytics produced from them are linked at the metadata level. By traversing these links, users can browse through a large network of analytical resources more effectively (Siow, Wang, and Tiropanis 2016). Once related resources are identified, users and applications can access them via the Web Observatory in a secure way.

Access control for sharing datasets The Web Observatory allows users to access datasets and analytics via a web interface. It also provides an API for applications to programmatically access analytical resources (Madaan et al. 2016; Siow, Wang, and Tiropanis 2016; Tinati et al. 2015; Tiropanis et al. 2014). The API is protected by OAuth 2.0[7] to ensure that access to any resource is controlled by the resource owner and can be delegated to applications without having to reveal the underlying credentials. Applications can also access multiple datasets simultaneously, through the API, to perform complementary analytics and information fusion. It is also possible to access live data streams and build real-time analytics (Madaan et al. 2016; Siow, Wang, and Tiropanis 2016; Tinati et al. 2015; Tiropanis et al. 2014). The Web Observatory lets users list or host datasets that are public or private. Access to private datasets is managed by the data owner, who hosts them on a Web Observatory node. Because access to datasets can be restricted, access to applications that make use of those datasets must be restricted as well (Tiropanis et al. 2013; Tiropanis et al. 2014). Resources can be queried using the Web Observatory API, which uses a JavaScript Object Notation (JSON) structured query language, and the mappings to the types of data stores are handled by the Web Observatory API and processed on the server side. This API acts as a secure middle layer between dataset locations and the end-user connections (Wang et al. 2017). The Web Observatory deploys an access control mechanism so that private datasets and analytics containing sensitive information are protected and accessible only to authorized users and applications.

Provenance for data quality The Web Observatory supports provenance of datasets and also their derivatives using the W3C PROV standard. This is achieved by adding provenance documents and linking datasets to provenance uniform resource identifiers in the Web Observatory administration interface. PROV-AQ[8] provides a pingback mechanism to discover provenance information and has been integrated into the Web Observatory infrastructure. This helps data publishers understand how the dataset has been used once it has been created and further supports usage tracking, transparency, and ultimately, users' control over how their datasets are used.

Terms and conditions for innovation on the Web Observatory   The PROSENT model (Wilson et al. 2016) considers whether the downstream use of datasets is in agreement with the data consent given by the data publishers on the Web Observatory platform. However, Wilson et al. (2016) also highlight the complexity of determining the consent violation and safeguarding the resources available on the Web Observatory owing to its distributed nature.

The Web Observatory and Data Governance Challenges

The distributed Web Observatory architecture is designed to address the governance challenges raised in the previous section. The important ingredient is the distributed nature of the resource, which allows data controllers to retain control, to determine how discoverable their data are, and ultimately to decide who gets access and under what conditions. In particular, we draw attention to the following aspects.

Privacy   The data owner's specified metadata description is available on the Web Observatory, and any access request is evaluated against the visibility of the resource. The data owners and publishers define partial or complete metadata for the discoverability of their resources on the Web Observatory platform.

Openness   Each observatory node defines its own sharing attributes on specific datasets. The Observatory uses reverse proxy, and the datasets are stored in the jurisdiction where they originated. The datasets available can be open or private as specified by the data owner or publisher.

Integrity   The provenance of analytical tools and datasets and the history of use provide a basis for trust for end users because they can see the derivations of their datasets, their transformation, and the context in which they have been used. Access patterns and usage tracking can be used to quantify the liability and accountability of stakeholders for data use, clarifying legal issues.

Sovereignty   The Observatory connects individual and organizational stores. It does not mandate that end users store the data within the Observatory node; rather, the datasets can be hosted at the end user's site.

Uncertainty   The support for access controls, provenance, and terms and conditions for innovation all help mitigate the uncertainty about liability and so forth. In the future, one could imagine the Web Observatory producing pro forma agreements for sharing and a set of best practices (cf. O'Hara 2019). However, given the international and cross-jurisdictional nature of the exercise, it is unlikely that this problem can be solved entirely by infrastructure.

Conclusion

As the cyber and real worlds become harder to treat separately with every new technology adoption, data are an increasingly important input to evidence-based policy making and to Internet governance questions that often arise because of the innovative ways people interact with the web (a 2019 project explores the infrastructural, technological, and legal issues involved with creating the Web Observatory to share data from the IoT[9]). This chapter describes how the diverse communities of academia, industry, and the public sector can be brought together to understand the evolution of the web and cocreate methodological evidence using the Web Observatory ecosystem to inform web and Internet governance policy development. The Web Observatory methodology itself, however, raises many governance issues of its own, and this chapter describes how its distinctive set of principles and freedom of data governance supports safe, ethical, and legal access to Observatory datasets and tools to support longitudinal data analysis and policy making across time, geopolitical boundaries, and topics.

Knowledge inferred from data in the Web Observatory about security, surveillance, and the promotion of state propaganda through social media platforms could provide evidence to policy makers to understand the effects of different national and supranational views of how the Internet should be controlled and developed. It has been argued that some governments are warping the Internet by attempting to align information flows with their jurisdictional boundaries (Chander and Lê 2014–2015; Mueller 2017; and see Mueller and Badiei's chapter 2). Beyond that, there is evidence of four different visions of the Internet coming to the fore, which are affecting its governance, driven by ideals such as openness, privacy, paternalism, and commercial imperatives. We also need to take account of government-sponsored hacking behavior occurring at scale, an ideological view that counterbalances the state-centric view (O'Hara and Hall 2018, 2020); these are the conflicting values mentioned specifically as a challenge in chapter 1. Resolving these conflicts is critical in the current fragmented scene of Internet governance, where we need to contrast national lenses with the multistakeholder view and with views such as that of the EU, which is using legislation such as the General Data Protection Regulation to protect individuals' online data while simultaneously projecting political power beyond its boundaries. Of course, the matter is complicated by the possibility that the Web Observatory itself requires certain ideological assumptions (for instance, openness) as a precondition for operating at all.

We have mentioned en passant a number of the challenges to Internet governance research set out in chapter 1; it should be clear by now that an architecture like the Web Observatory, by making the invisible visible through furnishing data, will help address most—although not all—of those challenges to some degree at least. We are left only with the concluding suggestion that the Web Observatory itself is one of the tools that need to be created, supported, and honed to enable Internet governance research, as cited in the final challenge. However, the Observatory is not only a tool; it demands a community of institutions prepared to make their data visible if not open, to cooperate in the research task, and to develop sufficient trust to facilitate the use and reuse of data. To that end, the status of the Web Observatory as a tool, as with any practicable technology, depends crucially both on architecture and on practice.

Acknowledgments

This work is supported by SOCIAM: The Theory and Practice of Social Machines. The SOCIAM Project is funded by the UK Engineering and Physical Sciences Research Council (EPSRC) under grant number EP/J017728/1 and comprises the Universities of Southampton, Oxford, and Edinburgh.

Notes

1. "Hypertext Transfer Protocol—HTTP/1.1," IETF Network Working Group, Request for Comments 2616, June 1999, https://www.w3.org/Protocols/rfc2616/rfc2616.html.

2. "Hypertext Markup Language—2.0," IETF Network Working Group, Request for Comments 1866, November 1995, https://tools.ietf.org/html/rfc1866.

3. "Web Observatory," accessed October 6, 2019, https://webobservatory.soton.ac.uk/.

4. "WSTNet Laboratories," WebScience Trust, accessed October 6, 2019, http://webscience.org/wstnet-laboratories/.

5. "Web Observatory Community Group," World Wide Web Consortium, accessed October 6, 2019, https://www.w3.org/community/webobservatory/.

6. "What Is a Reverse Proxy Server?," NGINX, accessed October 6, 2019, https://www.nginx.com/resources/glossary/reverse-proxy-server/.

7. "OAuth 2.0," OAuth.net, accessed October 6, 2019, https://oauth.net/2/.

8. "PROV-AQ: Provenance Access and Query," W3C Working Group Note, April 30, 2013, https://www.w3.org/TR/prov-aq/.

9. "IoT Observatory," Themes, PETRAS, accessed October 6, 2019, https://www.petrashub.org/portfolio-item/iot-observatory/.

References

Banyan. (2017, October 26). Is the world getting Myanmar wrong? *The Economist.* Retrieved from https://www.economist.com/news/asia/21730684-future-not-long-ago-deemed-bright-now-feels-bleak-world-getting-myanmar-wrong

Benkler, Y. (2006). *The wealth of networks: How social production transforms markets and freedom.* New Haven, CT: Yale University Press.

boyd, d. (2008). Facebook's privacy trainwreck: Exposure, invasion, and social convergence. *Convergence, 14*(1), 13–20.

Chander, A., & Lê, U. (2014–2015). Data nationalism. *Emory Law Journal, 64*(3), 677–739.

Davies, T. (2013, October 9). Web observatories: The governance dimensions [Blog post]. Retrieved from http://www.opendataimpacts.net/2013/10/web-observatories-the-governance-dimensions/

DeNardis, L. (2010). The emerging field of Internet governance. *Yale Information Society Project Working Paper Series.* Retrieved from https://papers.ssrn.com/sol3/papers.cfm?abstract_id=1678343

DeNardis, L. (2014). *The global war for Internet governance.* New Haven, CT: Yale University Press.

DeNardis, L., & Hackl, A. M. (2015). Internet governance by social media platforms. *Telecommunications Policy, 39*(9), 761–770.

De Roure, D. (2014, March 20). Web Observatories, e-research and the importance of collaboration. WST Webinar Series. Retrieved from https://www.slideshare.net/davidderoure/web-observatories-and-eresearch

Dutton, W. H., & Peltu, M. (2005). The emerging Internet governance mosaic: Connecting the pieces. *SSRN*. Retrieved from https://ssrn.com/abstract=1295330

Fry, L., Hall, W., Koronios, A., Mayer, W., O'Hara, K., Rowland-Campbell, A., Stumptner, M., Tinati, R., Thanassis, T., & Wang, X. (2015). Governance in the age of social machines: The web observatory. The Australia and New Zealand School of Government. Retrieved from https://eprints.soton.ac.uk/378417/

GDPR Individual rights. (n.d.). Retrieved December 6, 2017, from https://ico.org.uk /for-organisations/guide-to-the-general-data-protection-regulation-gdpr/individual -rights/

Hall, W., Hendler, J., & Staab, S. (2017). A manifesto for web science@ 10. *arXiv*. Retrieved from https://arxiv.org/abs/1702.08291

Hall, W., & Pesenti, J. (2017). *Growing the artificial intelligence industry in the UK*. Department for Digital, Culture, Media & Sport and Department for Business, Energy & Industrial Strategy. Part of the Industrial Strategy UK and the Commonwealth. Retrieved from https://www.gov.uk/government/publications/growing-the-artificial -intelligence-industry-in-the-uk

Hall, W., Tiropanis, T., Tinati, R., Booth, P., Gaskell, P., Hare, J., & Carr, L. (2013, May 1–3). *The Southampton University Web Observatory*. Paper presented at the First International Workshop on Building Web Observatories. *ACM Web Science*. Retrieved from https://eprints.soton.ac.uk/352287/

Hall, W., Tiropanis, T., Tinati, R., & Wang, X. (2016). Building a global network of web observatories to study the web: A case study in integrated health management. *Qatar Foundation Annual Research Conference Proceedings, 2016*(1), 3092.

Hall, W., Tiropanis, T., Tinati, R., Wang, X., Luczak-Rösch, M., & Simperl, E. (2014). The web science observatory: The challenges of analytics over distributed linked data infrastructures. *ECRIM News, 2014*(96), 29–30. Retrieved from https://eprints .soton.ac.uk/361437/1/ERCIM-69_p29-30.pdf

Hargittai, E. (2010). Digital na(t)ives? Variation in Internet skills and uses among members of the "net generation." *Sociological Inquiry, 80*(1), 92–113.

Hildebrandt, M. (2016). *Smart technologies and the end(s) of law novel entanglements of law and technology*. Cheltenham, UK: Edward Elgar.

Kittur, A., Chi, E., Pendleton, B. A., Suh, B., & Mytkowicz, T. (2007). Power of the few vs. wisdom of the crowd: Wikipedia and the rise of the bourgeoisie. *World Wide Web, 1*(2), 19.

Lanier, J. (2010). *You are not a gadget*. New York, NY: Knopf.

Madaan, A., Tiropanis, T., Srinivasa, S., & Hall, W. (2016). Observlets: Empowering analytical observations on Web Observatory. In *Proceedings of the 25th International*

Conference Companion on World Wide Web (pp. 775–780). Geneva, Switzerland: International World Wide Web Conferences Steering Committee. Retrieved from https://doi.org/10.1145/2872518.2890593

Mueller, M. (2017). *Will the Internet fragment?* Cambridge, UK: Polity Press.

Ngamassi, L., Ramakrishnan, T., & Rahman, S. (2016). Use of social media for disaster management: A prescriptive framework. *Journal of Organizational and End User Computing, 28*(3), 122–140. http://dx.doi.org/10.4018/JOEUC.2016070108

O'Hara, K. (2015). Data, legibility, creativity…and power. *IEEE Internet Computing, 19*(2), 88–91.

O'Hara, K. (2019). Data trusts: Ethics, architecture and governance for trustworthy data stewardship. Retrieved from http://dx.doi.org/10.5258/SOTON/WSI-WP001.

O'Hara, K., & Hall, W. (2018). *Four Internets: The geopolitics of Internet governance* (Paper no. 206). Centre for International Governance Innovation. Retrieved from https://www.cigionline.org/publications/four-internets-geopolitics-digital-governance

O'Hara, K., & Hall, W. (2020). Four Internets. *Communications of the ACM, 63*(3), 28–30. http://dx.doi.org/10.1145/3341722

O'Hara, K., & Shadbolt, N. (2015). The right to be forgotten: Its potential role in a coherent privacy regime. *European Data Protection Law Review, 1*(3), 178–189.

O'Hara, K., Shadbolt, N., & Hall, W. (2016). *A pragmatic approach to the right to be forgotten*. Global Commission on Internet Governance Paper Series (Paper no. 26). Centre for International Governance Innovation/Chatham House. Retrieved from https://www.cigionline.org/publications/pragmatic-approach-right-be-forgotten

Price, S., Hall, W., Earl, G., Tiropanis, T., Tinati, R., Wang, X., … & Groflin, A. (2017, April). Worldwide Universities Network (WUN) Web Observatory: Applying lessons from the web to transform the research data ecosystem. In *Proceedings of the 26th International Conference on World Wide Web Companion* (pp. 1665–1667). Geneva, Switzerland: International World Wide Web Conferences Steering Committee.

Shadbolt, N., O'Hara, K., De Roure, D., & Hall, W. (2019). *The theory and practice of social machines*. Cham, Switzerland: Springer.

Shneiderman, B. (2007). Web science: A provocative invitation to computer science. *Communications of the ACM, 50*(6), 25–27.

Singh, J., & Haziq, M. (2016). *Myanmar's Rohingya conflict: Foreign jihadi brewing*. S. Rajaratnam School of International Studies commentary CO16259. Retrieved from https://www.rsis.edu.sg/rsis-publication/icpvtr/co16259-myanmars-rohingya-conflict-foreign-jihadi-brewing/#.Xi7QmzKgLIU

Siow, E., Tiropanis, T., & Hall, W. (2016, October). *PIOTRe: Personal Internet of Things repository*. Poster presented at the International Semantic Web, Demos, Japan.

Siow, E., Wang, X., & Tiropanis, T. (2016, May). Facilitating data-driven innovation using VOICE observatory infrastructure. In *Proceedings of the Workshop on Data-Driven Innovation on the Web* (p. 5). New York, NY: ACM.

Stevens, D., & O'Hara, K. (2015). *The devil's long tail: Religious and other radicals in the Internet marketplace*. London, UK: Hurst.

Stieglitz, S., & Dang-Xuan, L. (2013). Social media and political communication: A social media analytics framework. *Social Network Analysis and Mining, 3*(4), 1277–1291.

Tinati, R., Wang, X., Tiropanis, T., & Hall, W. (2015). Building a real-time web observatory. *IEEE Internet Computing, 19*(6), 36–45.

Tiropanis, T., Hall, W., Hendler, J., & de Larrinaga, C. (2014). The Web Observatory: A middle layer for broad data. *Big Data, 2*(3), 129–133.

Tiropanis, T., Hall, W., Shadbolt, N., De Roure, D., Contractor, N., & Hendler, J. (2013). The web science observatory. *IEEE Intelligent Systems, 28*(2), 100–104.

Turkle, S. (2005). *The second self: Computers and the human spirit*. Cambridge, MA: MIT Press.

Twenge, J. M., Joiner, T. E., Rogers, M. L., & Martin, G. N. (2018). Increases in depressive symptoms, suicide-related outcomes, and suicide rates among US adolescents after 2010 and links to increased new media screen time. *Clinical Psychological Science, 6*(1), 3–17.

Vaidhyanathan, S. (2007). The Googlization of everything and the future of copyright. *University of California Davis Law Review, 40*, 1207–1231.

van Kleek, M., Smith, D. A., Tinati, R., O'Hara, K., Hall, W., & Shadbolt, N. R. (2014, April). 7 billion home telescopes: Observing social machines through personal data stores. In *Proceedings of the 23rd International Conference on World Wide Web* (pp. 915–920). New York, NY: ACM.

Verhulst, S., Noveck, B. S., Raines, J., & Declerq, A. (2014). *Innovations in global governance: Toward a distributed Internet governance ecosystem*. Global Commission on Internet Governance Paper Series (Paper no. 5). Centre for International Governance Innovation/Chatham House. Retrieved from https://www.cigionline.org/publications/innovations-global-governance-toward-distributed-internet-governance-ecosystem

Wagner, F. R. (2016, November). The Internet governance ecosystem: Where we are and the path ahead. In *Proceedings of the 22nd Brazilian Symposium on Multimedia and the Web* (pp. 5–6). New York, NY: ACM.

Wang, X., Madaan, A., Siow, E., & Tiropanis, T. (2017, April). Sharing databases on the web with Porter Proxy. In *Proceedings of the 26th International Conference on World Wide Web Companion* (pp. 1673–1676). Geneva, Switzerland: International World Wide Web Conferences Steering Committee.

Wilson, C., Tiropanis, T., Rowland-Campbell, A., & Fry, L. (2016). Ethical and legal support for innovation on web observatories. In *Proceedings of the Workshop on Data-Driven Innovation on the Web* (p. 1). New York, NY: ACM.

Wong, J. C., Safi, M., & Rahman, S. A. (2017, September 20). Facebook bans Rohingya group's posts as minority faces "ethnic cleansing." *The Guardian.* Retrieved from https://www.theguardian.com/technology/2017/sep/20/facebook-rohingya-muslims -myanmar

Zuboff, S. (2019). *The age of surveillance capitalism: The fight for a human future at the new frontier of power.* London, UK: Profile.

7 Taking the Growth of the Internet Seriously When Measuring Cybersecurity

Eric Jardine

No one seriously thinks of the Internet as a static system. By both design and effect, the Internet ecosystem exists in an inherently plastic and fluid state. In two very particular ways—in terms of both its scale and composition—the Internet has morphed from its early beginnings as a small set of networked research computers into something much bigger and far more diverse.

Each day, for example, thousands of new entrants connect to the Internet for the very first time. While year-over-year growth of the Internet-using population is slowing, it remains above 10 percent per annum (Meeker 2017, 6). With 3.7 billion users in 2016 (Internet World Stats 2017), this rapid pace of change resulted in potentially over 4 billion users in 2018. By dint of global demographics, many of these new users hail from developing-world countries and more autocratic regimes, potentially leading to growing levels of contention in the global Internet governance regime (Bradshaw et al. 2015).

Increasingly, the Internet is also losing its human face as it rapidly becomes predominately an interconnection of nonhuman devices. By dint of sheer numbers, the Internet of Things (IoT), as it is commonly known, is fundamentally changing the nature of the online environment. In 2014, for example, there were some 3.8 billion IoT devices (Gartner Research 2015). By 2016, that number was predicted to have increased to 6.39 billion. In 2015, Gartner Research (2015) forecasted that almost 21 billion IoT devices could be connected to the Internet by 2020. In 2020, Statista Research puts the number of IoT devices at 30.73 billion and offers a prediction of 74.44 billion devices online by 2025 (Statista 2020). Moreover, while the pace of new-user growth is slowing, the rate of new IoT interconnections is still in an acceleratory phase, exhibiting a classic S-shaped pattern of innovation diffusion that was earlier followed by Internet connectivity as a whole (Hampson and Jardine 2016; Rogers 2010).

These changes to the underlying user base and composition of the Internet are fundamental. The wider Internet governance literature has accommodated the changing Internet landscape well. For example, as detailed in chapter 3 by Mueller and Badiei, Internet governance—and literature on the governance of the Internet more generally—has undergone several distinct phases of work and focus since the early 1990s, with a big move toward securitization and concerns over privacy since 2010. Likewise, Cogburn in chapter 9 shows with a quantitative textual analysis of Internet Governance Forum (IGF) transcripts that the dominant themes at the IGF have changed over time. In 2006, the three dominant themes were "mobile," "youth," and "ICANN" (the Internet Corporation for Assigned Names and Numbers). In 2011, the focus had shifted toward more rights-based issues, with "human rights," "young people," and "freedom of expression" being the most expressed phrases. Lastly, in 2017, changes again occurred with "women," "cybersecurity," and "news" being the top phrases in the corpus of IGF text. Other research has also started to contemplate what will happen to Internet governance when all devices are interconnected (Van Eeten 2017) or the implications of the IoT for law and policy within the wider Internet governance regime (DeNardis and Raymond 2017). In short, coping with change is a big part of Internet governance practice, and it is well captured in many strands of the literature.

And yet, as I argue here, coping with ecosystem change at the level of cybersecurity measurement is less often done. When developing metrics to measure and assess the state of cybersecurity, the changing scale and composition of the Internet is often ignored. This error is implicit in the measures themselves. Two errors, in particular, are potentially distorting our sense of the state of cybersecurity. First, trends in security incidents as reported by information technology (IT) security vendors and governments tend to be depicted as absolute count (e.g., 1 million web-based attacks in 2016). This number fails to account for the significant increase in new users of the Internet (Jardine 2015, 2018). As the ecosystem grows, attacks, hacks, or vulnerabilities will also increase, everything else being equal. For measures of cybersecurity incidents to be meaningful, they need to be expressed as a rate, which requires that the absolute count of security incidents, say, be normalized around the size of the ecosystem, be it users, websites, data flows, or emails.

The second way in which current measures of cybersecurity are potentially misleading is a result of their often extreme level of aggregation (Jardine

2017). Devices and users of every kind are often grouped together as if they are effectively the same thing. The trouble here is that different parts of the ecosystem are growing faster than others, and moreover, the fastest-growing parts, such as the IoT, tend to be the most susceptible to being hacked (HP 2017). This pairing creates a very real danger of what is known as Simpson's Paradox, in which aggregate trends in security incidents show things getting worse while the disaggregated trends show the state of online security to be improving over time (Jardine 2017). Data from US data breaches can showcase the problem, but simulation results suggest that the phenomenon occurs in any situation in which groups are differentially vulnerable, and the fastest-growing group is the most vulnerable of the lot.

In the chapter's first section, I discuss trends in current cybersecurity metrics. I illustrate several potential perils and pitfalls in this section, too, and propose some cautionary steps that can help researchers produce the best possible metrics of cybersecurity. In the second section, I show that even the best cybersecurity measures will produce biased over-time trends if they do not take into account the ongoing growth Internet denominator, to use the mathematical term. The third section points to existing over-time trends in cybersecurity measures. It shows further how Simpson's Paradox (or aggregation bias, as it is also known) exists in data breach data and plausibly biases many aggregate indicators of insecurity. I conclude with thoughts on why accurate measurement of trends in cybersecurity matters and how cybersecurity researchers can approach their topic to form the best possible descriptive inferences.

What We Know about Measuring Cybersecurity

Measures of insecurity are never in short supply within the Internet ecosystem. Yet many such measures are hardly worth the bytes on which they are stored. Even the simple estimate of the total cost of cybercrime is hugely variable. One study by the Center for Strategic and International Studies and the antivirus vendor McAfee (2014), for example, estimated that the likely global cost of cybercrime was on the order of $400 billion. More recent estimates have ballooned these already significant numbers further still. A year after the center and McAfee produced their figure, Juniper Research (2015) estimated that the global costs of cybercrime could reach as high as $2 trillion per annum as early as 2019. Another report, by the firm Cybersecurity

Ventures, predicted that global cybercrime costs could actually have been as high as $3 trillion in 2015 and may well potentially increase by stunning proportions to as much as $6 trillion in 2021 (Morgan 2016).

Much of these costs are the result of an accumulation of a thousand small cuts. Each year, firms of all stripes and sizes are breached and their records are stolen or otherwise compromised. The Ponemon Institute conducts an annual survey of the cost of data breaches among hundreds of firms globally. As detailed in figures 7.1 and 7.2, both the average cost per breached record and the average total cost that a firm faces in the aftermath of a breach are undergoing a general rise.[1] In 2011, for example, the cost per breached record was $130. By 2017, the average breach cost had increased to $148 per record, amounting to a not inconsiderable increase of 13.9 percent. The total average cost that firms need to absorb in the fallout of a data breach is also increasing, rising from $3.13 million in 2011 to roughly $3.86 million in 2017 (Ponemon Institute 2013–2018).

These trends in the cost of data breaches are roughly mirrored by the movement in their frequency. Within the United States, the Privacy Rights Clearinghouse (2020) has collated an ongoing record of US data breaches from 2005 onward. Both the number of breaches and the summed total number of compromised records is increasing over time. The total number of US data breaches (from all threat vectors), for example, increased from

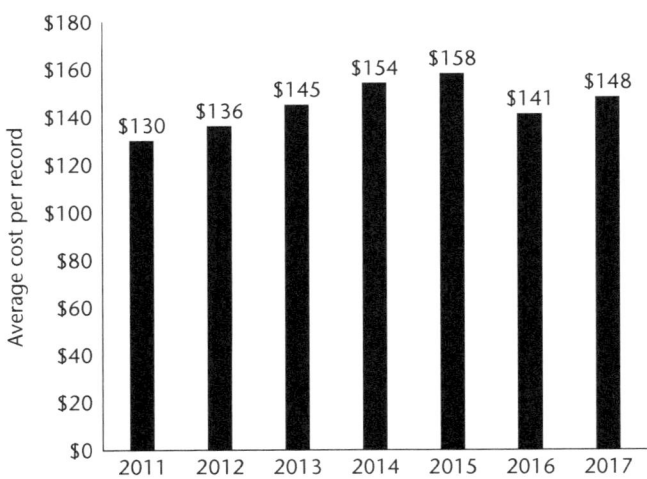

Figure 7.1
Ponemon Institute average cost per record breached.

Figure 7.2
Ponemon Institute global data breach costs.

136 in 2005 to 832 in 2016, amounting to a total increase of 512 percent. The total number of breached records, on the other hand, has increased at a far more prodigious rate. Here, the total number of compromised records[2] has increased from 55,101,241 in 2005 to 4,626,238,665 in 2016, which includes the 2013 megabreach of 1 billion Yahoo! users. This change amounts to an increase of 8,296 percent. (We return to these numbers in the third section, as the story they tell can be deceptive.)

These numbers suggest that one massive ongoing shift in the cybersecurity environment is toward more pronounced outcomes, what Nassim Taleb (2010, 33–34) would call the world of "extremistan." This notion is broadly in line with much of the growth in data breaches being accounted for by the existence of so-called heavy tails, or extreme outliers that drive a lot of the variation (Edwards, Hofmeyr, and Forrest 2016). Growing digitization, interconnectivity, and market concentration all affect cyber risk and help make the occurrence and fallout of a data breach far more pronounced (Geer, Jardine, and Leverett 2020).

These measures all point in the same approximate direction, indicating roughly that the state of cybersecurity is worsening. IT security practitioners have certainly internalized this message. The online Index of Cybersecurity (2020), for example, conducts a monthly survey of IT security professionals to gauge their concern over the state of the online environment. From a base value of 1,000 in March 2011, the index score has exhibited a decidedly worsening trajectory, moving to a score of 4,762 on July 4, 2019. These collective sentiments are broadly in line with the idea of former high-level policy makers who openly fret about the potentially severe consequences of a cyberattack. Former Homeland Security secretary Tom Ridge, for example, argued in late 2016, "Notwithstanding the pain and horror associated with a physical attack, ... the potential for physical, human, and psychic impact with a cyberattack, I think, is far more serious" (Roberts 2016). Clearly, the general sentiment is that efforts to improve cybersecurity outcomes are, so far at least, a losing battle.

At the same time, however, there are a number of telling reasons to suggest that we might know less than we think about the state of online security. The broad point of the literature is that most measures are, as is often the case, biased—and only a very select few useful. What this means in practice is that our view that cybersecurity is a mug's game might not be as well grounded in empirical reality as we might like.

Anderson and his colleagues, for example, produced a rigorous estimate of the cost of cybercrime, which is far less than the trillions upon trillions of dollars cited earlier. This research team started by disaggregating costs into defense costs, indirect losses, and the direct losses from cybercrime (2012, 5). The disaggregation of costs is an important step, as direct costs to victims or the benefits to attackers are often only a small portion of the overall social costs due to cybercrime. The difficulty is the working of network externalities, only in reverse (Anderson and Moore 2006). Rather than the benefits of network interconnection increasing at the pace of Metcalfe's Law (i.e., number of connected devices, n^2), the tendency is for the costs of cybercrime to increase disproportionately as network interconnection goes up. But a majority of these social costs do not translate into dollars in the bank accounts of malicious actors.

The point is well shown by the huge gap that can exist between what cybercriminals make and the destruction that they can bring. In one study of the individual economics of phishing campaigns (Herley and Florêncio 2009), the authors persuasively argue that many estimates of the high value of phishing campaigns for individual attackers (sometimes on the order of thousands of dollars per day) is likely woefully inaccurate. Since phishing attacks target a common pool resource (users' wallets) that cannot regenerate at an infinite pace, the phishing activity of multiple attackers is likely to drive down the benefits until each attacker is basically earning what their skills would return in the noncriminal world. In more concrete dollar terms, Herley and Florêncio (2009, 67) suggest that rather than earning thousands a day, as some accounts suggest, phishers in 2008 likely shared a $61 million pot some 113,000 ways, or roughly $540 per attacker. At the same time, the social cost of phishing campaigns is enormous, with billions being poured into defense and remediation.

Overall, these numbers suggest that estimates of the high-flying benefits of cybercrime might be greatly exaggerated and that there is likely an asymmetry, due to network externalities, between malicious actor benefits and the damage done. But troubling estimations are not confined simply to metrics for the monetary gain to malicious actors. The problems with correctly measuring cybersecurity indicators of various stripes are manifold.

One such example is the lack of a common measurement foundation from which conclusions can be drawn (Brecht and Nowey 2013). A good concrete instance of this problem of sample composition and metric design is to look again at the cost of data breach studies produced by the Ponemon Institute. The object of Ponemon's efforts are twofold: provide a static snapshot of what is going on in the ecosystem and provide an over-time measure to assess whether data breaches are getting better or worse. The Ponemon data breach studies are helpful on the first count but suffer from a shifting-sample problem on the second.

As detailed in table 7.1, the Ponemon results are based on a growing population of both countries (increasing from 8 in 2011 to 19 in 2017) and firms (from 199 in 2011 to 477 in 2017). The growing sample increases

Table 7.1
Ponemon Institute's shifting sample.

	Countries	Firms
2011	8	199
2012	9	277
2013	10	314
2014	11	350
2015	12	383
2016	13	419
2017	19	477

coverage. But this shifting-sample population can also introduce biases when looking at the trends in the data over time. If a newly added country is from a highly targeted part of the world, then the average will be pulled up, especially since data breach outcomes are highly power-law distributed and characterized by extreme variance (Edwards, Hofmeyr, and Forrest 2016). All the other countries could be performing the same or even better than before, but the new entrant to the sample makes the over-time trend appear as though things are worsening. Likewise, if the new firms that join the sample are highly targeted (firms from the finance or health care sectors, say), then the averages could again be biased upward. The point, really, is that with a shifting-sample population it is hard to say if an increase in the average cost per breached record is really due to an increased cost per firm or due to the inclusion of new actors whose average cost per record is higher than the rest of the population under question (more on this problem of aggregation in the third section).

In some ways, the law of large numbers—in which averages become stable over a large enough population size—suggests that the shifting-sample base might not have much effect. Randomly sampling (which is not really what is being done) from the population of firms should eventually lead to a stable sample mean and normal distribution. However, stable means are common only in a Gaussian, or bell-shaped, distribution. When the phenomenon that is being surveyed follows a more extreme distribution, such as an exponential or power-law distribution, then measures of average costs can become terribly misleading (Florêncio and Herley 2013). Indeed, when power laws are at work, extreme values of outliers in the sample can cause shifts in the mean so radical that the average becomes nearly meaningless as a guide for policy.[3] Median cost—which Ponemon does not, but probably should—use—is then a simple and far better estimation of a middling value.

A host of problems are added on to issues of incorrectly estimated costs, shifting samples, and extreme distributions. Many areas of the social world, for example, use voluntary reporting mechanisms to tally up who has been subjected to what and why. Surveys are a classic example. The problem here is that cybercrime surveys can be marred by a curious admixture of both under- and overreporting (Florêncio and Herley 2013; Kshetri 2006). For some, being the victim of a cybercrime is a traumatic event and one that people want to express out loud, in print or online, to whomever will listen. For others, incentives might align to discourage disclosure of a hack, which probably happens in countries lacking mandatory data breach disclosure regulations for firms (Laube and Böhme 2016). Additionally, a lack of cybersecurity loss data stemming from incidents complicates risk pricing, necessitating cleverly innovative (yet somewhat error-prone owing to extrinsic factors) actuarial models (Woods, Moore, and Simpson 2019).

Finally, it is more than possible that the production of cybersecurity metrics can be captured by vested interests. The most obvious potential set of culprits, but by no means the only ones, are IT security vendors. These companies are in the uncomfortable position of researching a topic and producing estimates about the state of cybersecurity while also selling a product the demand for which is based, at least in part, on perceptions of online insecurity. This tension does not always play out in favor of bias, but it often can. As Anderson and colleagues point out,

> One problem with Symantec is a conflict of interest: their reports are published principally for marketing reasons. It is not surprising, then, that sometimes their data are consistently over-reported. For example, we studied more closely one statistic measured in several reports, which tracks the proportion of malicious code that exploits confidential information. In volume 12 of the report, covering January to June 2007, 65% of malicious code exploits confidential information, compared to just 53% in the previous six months. However, the earlier report claimed 66% for this period of July to December 2006. (2008)

There are, clearly, a host of different methodological and pragmatic biases that can cloud our view of cybersecurity. However, while improved measurement techniques that account for missing evidence, shifting samples, extreme distribution types, under- and overreporting of statistics, and vested interests would all be a step in the right direction, even the best metrics as they currently stand will remain biased. The trouble here is that some fundamental changes to the Internet ecosystem—most notably its growing scale and shifting demographic constitution—have not been accounted for well enough in the formulation of cybersecurity metrics. In the next two sections, I detail the distorting effects that these oversights can have on our view of cybersecurity.

Expressing Cybersecurity Incidents as a Rate

In the analogue world, indicators of insecurity or crime are always expressed as a rate (for example, 1,000 armed robberies, murders, or burglaries per

100,000 people). Expressions of this sort are useful because they correct for the simple fact that a larger population size should see more activity, both good and bad (see Jardine 2015, 2018). In an ecosystem with a denominator that is growing as fast as the Internet in terms of users, interconnections, traffic volume, and more, failing to correct for the changing size of the system will bias even the best-devised metrics of cybersecurity. The publication of statistics like the count of breached records is, in other words, inherently distortive and misleading.

Indeed, if the denominator (i.e., a measure of the size of the ecosystem) is growing fast enough, then the divergence between counts (number of records breached) and rates (number of records breached per 100,000 records or Internet users) can be significant, often so much so that one set of numbers could actually indicate that the state of cybersecurity is getting worse while the other indicates that things are getting better.

Consider, as a useful case in point, a comparison of the count of phishing websites taken from reports by the Anti-Phishing Working Group (2008–2016) with that same count normalized around the number of websites (see Jardine 2018 for more details on this analysis). As detailed in figure 7.3, phishing websites per quarter are steadily increasing in number. Certainly, more attack vectors do not always mean more insecurity necessarily. And phishing websites are a good example of that principle. With low latency blacklisting and other technical protections built into most modern browsers, malicious websites are not likely to remain effective for very long and are far less likely now than in the past to be able to infect a passerby in a drive-by attack. That being said, for a constant level of technological protections, more threats should roughly translate into more insecurity, so leaving

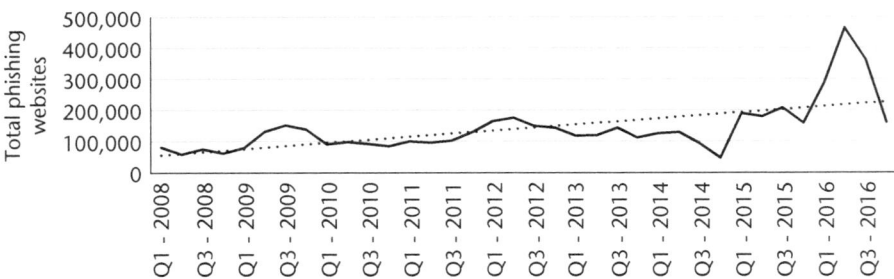

Figure 7.3
Observed phishing websites.

aside potential remedial countermeasures for a moment, let's just assume that more phishing websites could mean more insecurity online. Under that assumption, figure 7.3 clearly shows a worsening state of online security.

A starkly different picture emerges if you look, not at the count of the total number of observed phishing websites, but instead at per year rate of phishing websites per 100 websites or per 1 million Google searches.[4] These two denominators measure two separate aspects that help contextualize trends in the number of phishing websites. Obviously, phishing websites exist within a pool of all websites, so they are some proportion of the whole, just as the victims of crimes are some proportion of the whole population in a given area. Google searches are a measure of online activity. While Google's PageRank algorithm ensures that people are not simply wandering around randomly online, Google search activity could capture the amount of time people are spending online and so the amount of time within which they might stumble across a malicious website.

Counts and rates, in this case, tell very different stories. As shown in figure 7.4, rather than seeing a worsening situation as was the case with more phishing websites, both normalized measures of the rate of phishing websites are in decline. This declining rate suggests a couple of things. First, it means that the number of websites or Google searches is growing

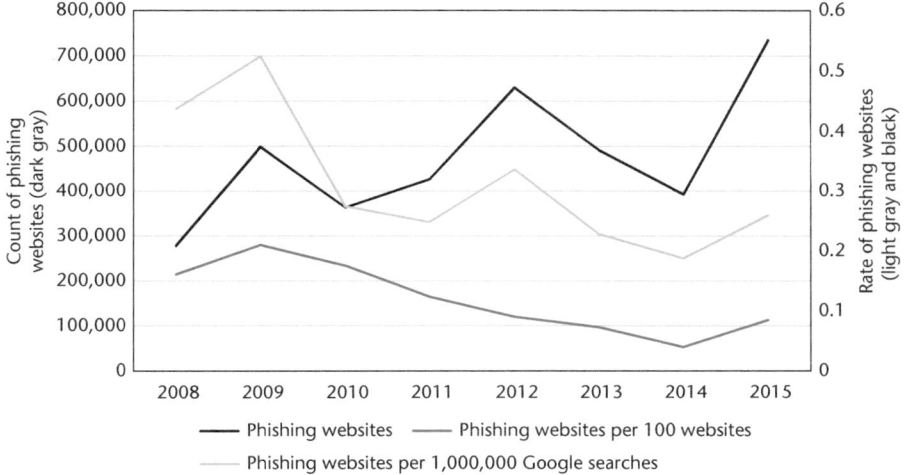

Figure 7.4
Counts versus rates of phishing websites.

faster than the number of malicious websites. This differential growth rate effectively entails that the population within which phishing websites live (all websites) or people's time online (Google searches) are outpacing the growth of phishing websites. Second, framed in propensity-based terms, these differential growth rates mean that it is increasingly *much less likely* that an average person doing average things online is going to encounter a malicious website. In this very important sense, using rates that capture the growing size of the Internet ecosystem when measuring cybersecurity tells a story of improved security.

Rates may or may not say that online security is getting better—if they deterministically did, then something is wrong with the measure. But, by simple dint of the mathematics involved, so long as the Internet ecosystem continues to grow, the rate will always paint a better picture than the absolute count of incidents, vulnerabilities, or costs. Indeed, as I have shown before (Jardine 2015), when the Internet is getting bigger, rates will favorably contrast with counts for year-over-year comparisons in one of the following three ways:

- The count says things are getting worse, but the rate says cybersecurity is getting better.
- The count says things are getting better, but the rate says cybersecurity is improving even faster.
- The count says things are getting worse, but the rate says that cybersecurity is getting worse at a slower pace.

These differences are crucial, as they imply that a year-over-year count of zero-day exploits, security incidents, phishing attacks, and so on will be biased. Using the best-designed numerator measure in the world—a design that is challenging in its own right—helps avoid some of the problems sketched out in the first section, but a failure to normalize these very same metrics around the growing size of the Internet ecosystem will result in a flawed measure that is not only potentially exaggerating the poor state of cybersecurity but that may well say everything is getting worse when it is actually getting better.

Aggregation and Bias in Cybersecurity Measures

There is, too, another way in which cybersecurity researchers may fail to adequately grapple with the idea that the Internet is a dynamic system that is growing and changing right under our feet. In this case, the pressing problem is the level of aggregation of most publicly available cybersecurity data. When data from various groups are aggregated, the results can fall prey to what is known as Simpson's Paradox (Simpson 1951). When Simpson's Paradox is at work, aggregate trends that show a worsening state of cybersecurity can actually be based on underlying data that show the state of cybersecurity to be improving year-over-year (Jardine 2017). Overaggregation, in other words, can produce bias.

Unfortunately, a lot of currently available trend data in cybersecurity exist at just such an extreme level of aggregation. IT security software vendors, such as Kaspersky Labs or Symantec, produce annual reports detailing the volume of threats that they contend with each year. While the data are a bit noisy, the general trend line observed by both companies is decidedly positive, tending toward more and more attacks over time.

Figure 7.5, for example, plots two forms of online attacks as observed by Kaspersky and Symantec. In Kaspersky's case, its annual Security Bulletin (2008–2016) contains a count of the total number of attacks launched from web-based sources, which would exclude some forms of malicious activity such as insider threats or USB-based malware. Symantec, for its part, includes a count for the number of blocked attacks in its annual Internet Security Threat Report (2013–2016).

In both cases, observed attacks are generally going up over time. Kaspersky's measure, for instance, has increased from 23,680,646 attacks in 2008

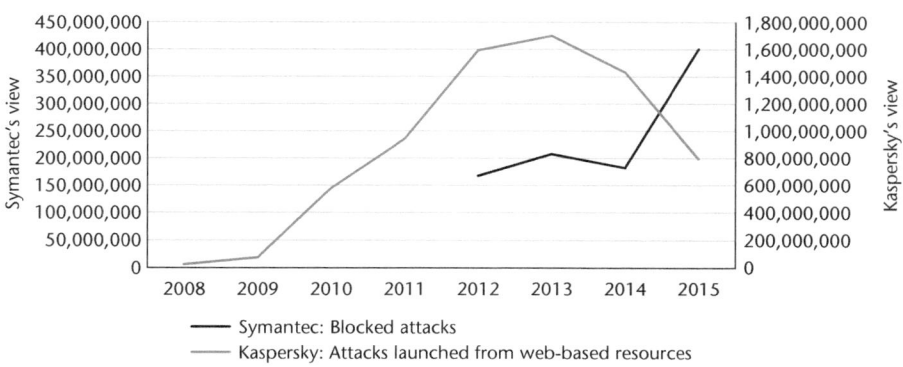

Figure 7.5
Aggregate attacks over time.

to 798,113,087 attacks in 2015, amounting to a growth in attacks of some 3,270 percent. Symantec's blocked attacks have likewise increased, by some 139 percent, growing from 167,900,000 blocked attacks in 2012 to 401,500,000 attacks in 2015. Obviously, the former point about the perils of count measures (as opposed to rates) suggest that these worsening trends might not be as bad as they seem. But these trends might be misleading for an additional reason: they fail to factor in the changing face of the Internet ecosystem. These statistics pay no attention to the type of devices being targeted, the baseline vulnerability of different user groups, the sectors within which these devices are housed, or even the region of the world where the devices reside. Aggregating these differences is potentially a major source of bias.

Take another example of seemingly bad overall trends, where sufficiently disaggregated data allow for a more fine-grained analysis of the problem of aggregation. The Privacy Rights Clearinghouse provides a record of publicly known data breaches in the United States. Asking what the trend is in data breaches and in the number of breached records shows the problem of overaggregation. Figure 7.6 shows the trend in the number of disclosed data breaches that resulted from all recorded attack vectors. There are generally more data breaches now than there were in the middle of the first decade of the 2000s. Figure 7.7 does the same for both the total count of breached

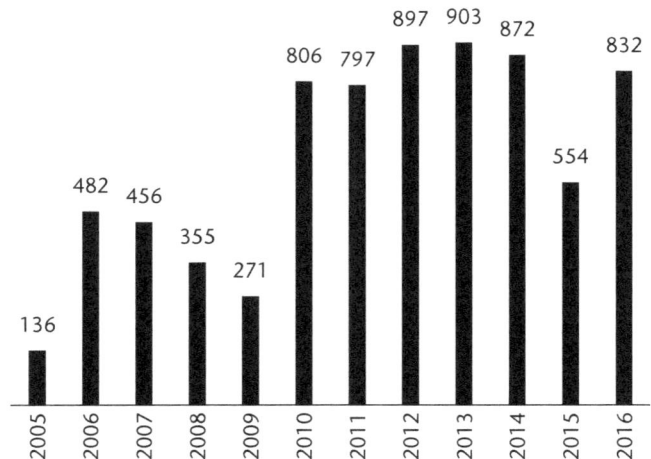

Figure 7.6
Total number of breaches.

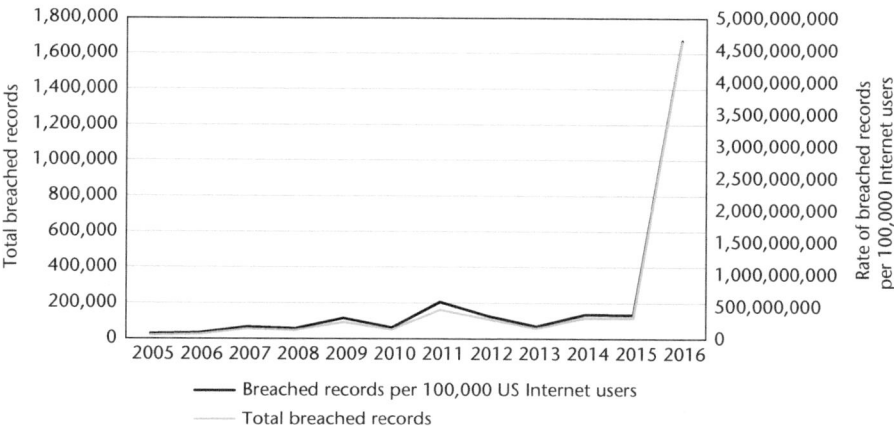

Figure 7.7
Total counts and rates of number of breached records.

records and the rate of breached records per one hundred thousand US-based Internet users. The trends track one another and are decidedly positive. From these three aggregate figures, whether the count of breaches or counts or rates of the number of compromised records, the primary conclusion to be drawn seems to be that the trend in cybersecurity is heading in a worsening direction.

But to its credit, the Privacy Rights Clearinghouse also provides a record of attacks disaggregated by sector. This disaggregation allows an analysis of over-time trends at a finer-level detail. In other words, it allows a sense of what is happening in specific parts of the economy, not just overall. Using data from the Privacy Rights Clearinghouse (2020), Table 7.2 highlights what is happening in the aggregated trend in the number of data breaches, the count of breached records, and the rate of breached records per one hundred thousand US Internet users. It also shows what is happening in each disaggregated sector.

Each aggregate measure tells the same story: cybersecurity is getting worse. The disaggregated story is far more nuanced and often the opposite. When counting the total number of breaches, the aggregate trend is positive, but only three (other business, retail, and health care) experienced more breaches in 2005–2016. In short, the total count of data breaches per year is declining in the majority of sectors. Likewise, the overall trend in the number of breached records is positive, and is similarly positive for all but

Table 7.2

Aggregate versus disaggregate breach data, 2005–2016.

Trend direction	Aggregate	Financial services	Other business	Retail	Education	Gov/mil	Health care	Nongovernmental
Total breaches	Positive	Negative	Positive	Positive	Negative	Negative	Positive	Negative
Counts of records	Positive	Positive	Positive	Positive	Negative	Positive	Positive	Negative
Breached records per 100,000 US Internet users	Positive	Negative	Positive	Positive	Negative	Negative	Positive	Negative

the educational and nonprofit sectors. But as discussed in the previous section, counts of compromised records are deceptive and are best normalized around the changing size of the Internet ecosystem. Doing so with a crude measure of the number of US-based Internet users taken from World Bank Indicators (a proxy for the extent of digitization) still produces a positive overall trend in the rate of breached records. The disaggregate story is radically different, however. Now, four of the seven categories exhibit a negative trend in the breached records rate per one hundred thousand Internet users from 2005 to 2016.

The point is that the level of aggregation in the data can decidedly affect the view that you get. Further disaggregation within each sector would likely review additional nuances. Grouping firms on the basis of their level of information security sophistication would be even more revealing. These results are also not confined to data breaches in which some semblance of sufficiently disaggregated data allow an illustration of the problem. In fact, this problem of Simpson's Paradox, or aggregation bias, emerges whenever three simple conditions hold (Jardine 2017):

- The online population can be divided into groups.

- Some of these groups are more vulnerable than others.

- Each group's rate of growth mirrors its vulnerability, with the most susceptible group growing the fastest.

To phrase these conditions a bit differently, if groups of online users or devices can be rank ordered in terms of vulnerability and the most vulnerable groups grow faster than the less vulnerable groups, then extensive simulation results in both normal and power-law-distributed settings indicate that aggregate trends will show a worsening situation, even as the disaggregate results all show security improving (Jardine 2017). The massive growth rate of highly vulnerable IoT devices almost ensures that overaggregation is affecting cybersecurity statistics today, making things seem worse than they really are. More generally, if a get-to-market-first-and-do-security-later attitude characterizes the future ethos of technology expansion (as it has in the past), then Simpson's Paradox is likely to be a persistent issue for cybersecurity metrics. Aggregate trend data, in other words, that fail to take into account the changing composition of the Internet are likely to be biased, because the focus on the forest misses all the trees.

Conclusions for Policy and Research

This chapter advances a simple idea. While cybersecurity researchers are highly attuned to some discrete changes in the Internet ecosystem (shifting malware variants, for example), many of our current metrics tend to miss the forest for the trees by ignoring the very real measurement implications of an ever-changing online world. In this chapter, I show that a failure to recognize these fundamental shifts in the metrics we use can produce measurement bias in two ways: via a lack of normalization and through overaggregation.

Drawing on some of my past research (Jardine 2015, 2018), I have shown here how using a count measure to express insecurity introduces bias in year-over-year trends by failing to grapple with the simple fact that a growing Internet should have more malicious activity. Instead, I argue that cybersecurity researchers need to adopt an age-old method from the analogue world and start normalizing cybercrime statistics around the growing Internet. The Anti-Phishing Working Group, whose phishing website data I present here, could usefully begin normalizing its counts to great effect— assuming a good measure of an agreed-on denominator could be found.

Of course, the most effective denominator for each measure of insecurity is still an open question. The number of Internet users offers a simple generic measure that will suffice in many cases. Yet in some discrete instances, such as zero-day exploits, a better denominator might be either lines of code (if we knew that number) or computer programs. Zero-day exploits in the code of mobile applications, for example, could be normalized around the total number of mobile applications to produce a better measure than just a count of vulnerabilities. For phishing websites, the population of domains, search traffic, or websites might work best in other cases, as discussed earlier.

I also show that more attention needs to be paid to the changing face of the Internet ecosystem in the process of devising cybersecurity metrics and collecting and analyzing data. Overly aggregate measures can point in one direction, while disaggregated numbers can show another story altogether. This phenomenon of Simpson's Paradox is quite general and likely holds whenever groups have different levels of vulnerability and growth rates (Jardine 2017). Data breach data from the United States exemplifies the potential problem. On the basis of total counts of breached records, for example, a policy maker, regulator, or corporate C-suite executive might decide that more needs to be done to improve cybersecurity. But the disaggregate trend in that firm's sector might show an improving scenario, suggesting that current measures and levels of investment are sufficient. All policies and investments have opportunity costs. Missing out by spending too much on security can reduce the usability of services and forgo other, potentially more beneficial activities.

The results of this analysis are important for both policy makers (broadly defined) and Internet governance researchers.

For policy makers, the results of the analysis suggest that current cybersecurity trends might be misleading. While more and better data would be needed to say for sure, the results presented earlier often indicate an improvement, rather than worsening, of the state of cybersecurity in both phishing websites and US-based data breaches across many discrete sectors of the economy. Data ideally inform public policy development, as bad policy can cause a host of unintended consequences and waste resources countering problems that are only partially real. Good policy follows only from good metrics, however, and this chapter proposes two discrete ways in which stakeholders, such as governments, computer security incident response teams, and security vendors, can collect more meaningful data.

For researchers, the analysis speaks to a few practices that ought to be adopted during data collection and analysis: normalize counts and, as much as possible, disaggregate data to control for potential so-called lurking confounders that give rise to Simpson's Paradox. Beyond that, the research prescriptions also speak to many of the core themes of this research compendium volume. Proprietary or closed systems, for example, diminish the transparency of collected data. Such a lack of transparency makes assessing data quality more difficult, but it also makes research-initiated corrections difficult to do after the fact. Large (big) data—an increasingly common feature of today's online environment—are potentially a boon, since they can include sufficient information to properly normalize incident counts and disaggregate trends to sufficiently fine-grained levels. Last, growing market concentration and increasing technical complexity are persistent trends online that directly affect cybersecurity risk (Geer, Jardine, and Leverett 2020) and promise to continue the constant ebb and flow of the scale and composition of the Internet ecosystem in a way that will, undoubtedly, affect cybersecurity measures. Researchers need to be ever vigilant to ongoing changes online, as good policy and research requires good measures and data.

Notes

1. The Ponemon Institute studies are published the year after the year in which the data are collected. For example, the data in the 2016 report is typically for 2015.

2. The Privacy Rights Clearinghouse draws a distinction between a breached record involving financial data and records involving nonsensitive material such as emails. I use the total number of records here. These breaches are also only in deliberately malicious attacks (not from accidental disclosures or in unknown attacks) and with a known target.

3. In a power-law probability distribution, the smaller the alpha, the more pronounced the instability in the mean (see Neumann 2015).

4. The per 100 and per 1,000,000 designators are used to make the figure intelligible. They have no material effect on the analysis.

References

Anderson, R., Barton, C., Bohme, R., Clayton, R., van Eeten, M. J. G., Levi, M., Moore, T., & Savage, S. (2012). *Measuring the cost of cybercrime*. WEIS 2012. Retrieved from http://www.econinfosec.org/archive/weis2012/papers/Anderson_WEIS2012.pdf

Anderson, R., Bohme, R., Clayton, R., & Moore, T. (2008). *Security economics and European policy*. The Workshop on the Economics of Information Security, WEIS 2008. Retrieved from http://www.econinfosec.org/archive/weis2008/papers/MooreSecurity.pdf

Anderson, R., & Moore, T. (2006). The economics of information security. *Science, 314*, 610–613. Retrieved from http://tylermoore.ens.utulsa.edu/science-econ.pdf

Anti-Phishing Working Group. (2008–2016). *Phishing attack trends reports*. Retrieved from http://www.antiphishing.org/resources/apwg-reports/

Bradshaw, S., DeNardis, L., Hampson, F. O., Jardine, E., & Raymond, M. (2015). *The emergence of contention in global Internet governance*. Global Commission on Internet Governance Paper Series (Paper no. 17). Centre for International Governance Innovation/Chatham House. Retrieved from https://www.cigionline.org/sites/default/files/no17.pdf

Brecht, M., & Nowey, T. (2013). A closer look at information security costs. In R. Böhme (Ed.), *The economics of information security and privacy*. Berlin, Germany: Spring Science & Business Media.

Center for Strategic and International Studies (CSIS) & McAfee. 2014. *Net losses: Estimating the global cost of cybercrime*. Centre for Strategic and International Studies. Retrieved from https://csis-prod.s3.amazonaws.com/s3fs-public/legacy_files/files/attachments/140609_rp_economic_impact_cybercrime_report.pdf

DeNardis, L., & Raymond, M. (2017). The Internet of Things as a global policy frontier. *UC Davis Law Review, 51*(2), 475–497.

Edwards, B., Hofmeyr, S., & Forrest, S. (2016). Hype and heavy tails: A closer look at data breaches. *Journal of Cybersecurity, 2*(1), 3–14. Retrieved from https://doi.org/10.1093/cybsec/tyw003

Florêncio, D., & Herley, C. (2013). Sex, lies and cyber-crime surveys. In B. Schneier (Ed.), *Economics of information security and privacy III* (pp. 35–53). New York, NY: Springer. Retrieved from https://link.springer.com/chapter/10.1007/978-1-4614-1981-5_3

Gartner Research. (2015). *Gartner says 6.4 billion connected "things" will be in use in 2016, up 30 percent from 2015.* Retrieved from http://www.gartner.com/newsroom/id/3165317

Geer, D., Jardine, E., & Leverett, E. (2020). On Market Concentration and Cybersecurity Risk. *Journal of Cyber Policy, 5*(1).

Hampson, F., & Jardine, E. (2016). *Look who's watching: Surveillance, treachery, and trust online.* Waterloo, Canada: Centre for International Governance Innovation Press.

Herley, C., & Florêncio, D. (2009). A profitless endeavor: Phishing as tragedy of the commons. *NSPW '08 proceedings of the 2008 New Security Paradigms Workshop* (pp. 59–70). Retrieved from http://dl.acm.org/citation.cfm?id=1595686

HP. (2017). *HP study reveals 70 percent of Internet of Things devices vulnerable to attack.* Retrieved from http://www8.hp.com/us/en/hp-news/press-release.html?id=1744676#.WSMGKmjysuU

Index of Cybersecurity. (2020). Retrieved from https://wp.nyu.edu/awm1/

Internet World Stats. (2017). *Internet usage statistics: The Internet big picture.* Retrieved from https://www.internetworldstats.com/stats.htm

Jardine, E. (2015). *Global cyberspace is safer than you think: Real trends in cybercrime.* Global Commission on Internet Governance Paper Series (Paper no. 16). Centre for International Governance Innovation/Chatham House. Retrieved from https://www.cigionline.org/sites/default/files/no16_web_0.pdf

Jardine, E. (2017). *Sometimes three rights really do make a wrong: Measuring cybersecurity and Simpson's paradox.* Paper presented at the 16th Workshop on the Economics of Information Security, WEIS 2017. Retrieved from http://weis2017.econinfosec.org/wp-content/uploads/sites/3/2017/05/WEIS_2017_paper_18.pdf

Jardine, E. (2018). Mind the denominator: Towards a better measurement system for cybersecurity. *Journal of Cyber Policy, 3*(1), 116–139.

Juniper Research. (2015). *Cybercrime will cost businesses over $2 trillion by 2019.* Retrieved from https://www.juniperresearch.com/press/press-releases/cybercrime-cost-businesses-over-2trillion

Kaspersky Labs. (2008). *Kaspersky security bulletin 2008*. Retrieved from https://securelist.com/analysis/kaspersky-security-bulletin/36241/kaspersky-security-bulletin-statistics-2008/

Kaspersky Labs. (2009). *Kaspersky security bulletin 2009*. Retrieved from https://securelist.com/analysis/kaspersky-security-bulletin/36284/kaspersky-security-bulletin-2009-statistics-2009/

Kaspersky Labs. (2010). *Kaspersky security bulletin 2010*. Retrieved from https://securelist.com/analysis/kaspersky-security-bulletin/36345/kaspersky-security-bulletin-2010-statistics-2010/

Kaspersky Labs. (2011). *Kaspersky security bulletin 2011*. Retrieved from https://securelist.com/analysis/kaspersky-security-bulletin/36344/kaspersky-security-bulletin-statistics-2011/

Kaspersky Labs. (2012). *Kaspersky security bulletin 2012*. Retrieved from https://securelist.com/analysis/kaspersky-security-bulletin/36703/kaspersky-security-bulletin-2012-the-overall-statistics-for-2012/

Kaspersky Labs. (2013). *Kaspersky security bulletin 2013*. Retrieved from http://media.kaspersky.com/pdf/ksb_2013_en.pdf

Kaspersky Labs. (2014). *Kaspersky security bulletin 2014*. Retrieved from https://securelist.com/files/2014/12/Kaspersky-Security-Bulletin-2014-EN.pdf

Kaspersky Labs. (2015). *Kaspersky security bulletin 2015*. Retrieved from https://securelist.com/analysis/kaspersky-security-bulletin/73038/kaspersky-security-bulletin-2015-overall-statistics-for-2015/

Kaspersky Labs. (2016). *Kaspersky security bulletin 2016*. Retrieved from https://securelist.com/analysis/kaspersky-security-bulletin/72771/kaspersky-security-bulletin-2016-predictions/

Kshetri, N. (2006). The simple economics of cybercrimes. *IEEE: Security & Privacy, 4*(1), 33–39.

Laube, S., & Böhme, R. (2016). The economics of mandatory security breach reporting to authorities. *Journal of Cybersecurity, 2*(1), 29–41.

Meeker, M. (2017). Internet trends, 2017—code conference. *Kleiner Perkins*. Retrieved from http://dq756f9pzlyr3.cloudfront.net/file/Internet+Trends+2017+Report.pdf

Morgan, S. (2016). Hackerpocalypse: A cybercrime revelation. *Cybersecurity Ventures*. Retrieved from https://www.herjavecgroup.com/hackerpocalypse-cybercrime-report/

Neumann, J. (2015). Power laws in venture. *Reaction Wheel*. Retrieved from http://reactionwheel.net/2015/06/power-laws-in-venture.html

Ponemon Institute. (2013). *2013 cost of data breach study: Global analysis*. Retrieved from https://www.ponemon.org/local/upload/file/2013%20Report%20GLOBAL%20 CODB%20FINAL%205-2.pdf

Ponemon Institute. (2014). *2014 cost of data breach study: Global analysis*. Retrieved from https://www-935.ibm.com/services/multimedia/SEL03027USEN_Poneman_2014 _Cost_of_Data_Breach_Study.pdf

Ponemon Institute. (2015). *2015 cost of data breach study: Global analysis*. Retrieved from https://nhlearningsolutions.com/Portals/0/Documents/2015-Cost-of-Data-Breach -Study.PDF

Ponemon Institute. (2016). *2016 cost of data breach study: Global analysis*. Retrieved from https://www.ibm.com/downloads/cas/7VMK5DV6

Ponemon Institute. (2017). *2017 cost of data breach study: Global overview*. Retrieved from https://www.ibm.com/downloads/cas/ZYKLN2E3

Ponemon Institute. (2018). *2018 cost of data breach study: A global overview*. Retrieved from https://www.ibm.com/downloads/cas/AEJYBPWA

Privacy Rights Clearinghouse (2020). Data breaches. Retrieved from https://privacyrights .org/data-breaches

Roberts, D. (2016). Tom Ridge: Cyber attacks are now worse than physical attacks. *Yahoo! Finance*. Retrieved from http://finance.yahoo.com/news/tom-ridge-cybersecurity -attacks-are-now-worse-than-physical-attacks-170426390.html?soc_src=social-sh &soc_trk=tw

Rogers, E. M. (2010), *Diffusion of innovations*. New York, NY: Simon & Schuster.

Simpson, E. H. (1951). The interpretation of interaction in contingency tables. *Journal of the Royal Statistical Society, Series B, 13*(2). 238–241.

Statista. (2020). Internet of Things (IoT) connected devices installed base worldwide from 2015 to 2025. Retrieved from https://www.statista.com/statistics/471264/iot -number-of-connected-devices-worldwide/

Symantec. (2013). *2013 Internet security threat report*. Retrieved from http://www .symantec.com/content/en/us/enterprise/other_resources/b-istr_main_report _v18_2012_21291018.en-us.pdf

Symantec. (2014). *2014 Internet security threat report*. Retrieved from http://www .symantec.com/content/en/us/enterprise/other_resources/b-istr_main_report _v19_21291018.en-us.pdf

Symantec. (2015). *2015 Internet security threat report*. Retrieved from https://www .symantec.com/content/en/us/enterprise/other_resources/21347933_GA_RPT-internet -security-threat-report-volume-20-2015.pdf

Symantec. (2016). *2016 Internet security threat report*. Retrieved from https://www.symantec.com/security-center/threat-report

Taleb, N. (2010). *The black swam: The impact of the highly improbable*. New York, NY: Random House Trade Paperbacks.

van Eeten, M. J. (2017). Patching security governance: An empirical view of emergent governance mechanisms for cybersecurity. *Digital Policy, Regulation and Governance, 19*(6), 429–448.

Woods, D., Moore, T., & Simpson, A. (2019, June). *The county fair cyber loss distribution: Drawing inferences from insurance prices*. Paper presented at the 18th Workshop on the Economics of Information Security, WEIS 2019, Boston, MA. Retrieved from https://www.researchgate.net/publication/332861796_The_County_Fair_Cyber_Loss_Distribution_Drawing_Inferences_from_Insurance_Prices

8 Researching Technology Elites: Lessons Learned from Data Collection at Google and Facebook

Rikke Frank Jørgensen

This chapter reflects on methodological lessons learned from doing empirical research on Google and Facebook as part of a research project conducted 2015–2017 on the commercialized public sphere (Jørgensen 2017, 2018; Jørgensen and Desai 2017). The project set out to explore how the two companies frame human rights such as freedom of expression and privacy in relation to their platforms; how this framing is reflected at governance level (e.g., content regulation), contractual level (e.g., terms of service), and technical level (e.g., default settings); and the human rights implications of these practices. My motivation for researching these questions was twofold. First, the companies are increasingly powerful and effectively influence the boundaries for how billions of users may exercise human rights such as the right to privacy and freedom of expression. The human rights impacts of their practices may arise in a number of situations and decisions related to how the companies respond to requests from the government to restrict content or access user information; how they adopt and enforce terms of service; the design and engineering choices that implicate security and privacy; and decisions to provide or terminate services in a particular market (Kaye 2016, para. 11). Second, researchers have little insight into the norms and narratives that shape these companies' practices. As emphasized by platform scholarship (Ananny and Gillespie 2016), these systems, people, and values are too often out of reach of scholars whose job is "to make explicit, orderly, consistent—and open to critical analysis—these 'orientations' that are usually taken for granted by empirical researchers" (Calhoun 1995, 5).

The research draws on a science and technology studies approach, highlighting how the design and implementation of any particular technology are patterned by a range of social, political, and economic factors (Williams

1996)—for example, how a particular economic interest patterns the privacy features that a platform offers (see also Musiani's science and technology studies approach to Internet governance in chapter 4). My analytical
vantage point has been to think of discourses—or what Mansell (2013) and
Flichy (2007) call imaginaries—as an integral part of understanding these
specific platforms. Similarly to a paradigm, much of the knowledge generated by a discourse comes to form common sense (Edwards 1997, 40). It
is a "background of assumptions and agreements about how reality is to
be interpreted and expressed, supported by paradigmatic metaphors, techniques, and technologies and potentially embodied in social institutions"
(34). A discourse continually expands its own scope, occupying and inte-
grating conceptual space in a kind of discursive imperialism. Presuming
that discourses shape social practices and expectations, unpacking people's
way of framing and ascribing meaning to specific issues may help us understand and confront their common sense, underlying assumptions, and met-
aphorical point of reference (Jørgensen 2013, 6). In short, discursive frames
are used to situate events, fashion a shared understanding of the world, and
guide problem-solving (Barnett and Finnemore 2004, 33). Engaging with
representations such as connecting the world, building social infrastructure,
organizing the world's information, and making information universally
accessible can help unpack the relationship between a technology and its
embedded norms and values, as well as illuminate how a particular framing
benefits certain interests and downplays others (see also Hofmann's take on
discourse analysis in relation to Internet governance and multistakeholderism in chapter 12). In explaining why particular frames are promoted
over others, the project relies on an actor-oriented perspective (Long 2001),
emphasizing the agency of company staff and their meaning making, motivations, and strategies. This implies attention to how individuals explain
particular meanings and how these meanings are translated into practice.

In terms of data collection, I have relied on a context-oriented qualitative approach, using interviews and online material as key sources of data
(Huberman and Miles 1994). I collected data via interviews with Google
and Facebook staff members (primarily in Europe and the United States),
as well as via publicly available material, including public presentations by
company staff. To retain some level of privacy for interviewees, while quali-
fying the findings to the widest extent possible, I identify them by affiliation
but not by name.

In the following, I use the research case to discuss challenges pertaining to this type of policy-engaged qualitative research, focusing on *getting access* to staff within the organizations, the *interview situation* itself, and the *data analysis*. While the case and its analysis relate specifically to Google and Facebook, the chapter raises broader issues on concepts, methods, and frameworks for conducting Internet governance research. In the concluding section, I draw out some of the lessons learned and relate these observations to Internet governance as a research field.

Getting Access

Gaining access to interlocutors within Facebook and Google proved to be a major challenge in relation to the data collection. Whereas both companies had initially—via staff based in Denmark—agreed to participate in the research project, the subsequent process of locating people willing to talk about human rights turned out to be very difficult. Practical obstacles varied from lack of publicly available contact details to lack of response to emails. Neither company, for example, has contact details of staff listed on the company websites. In contrast, many of the employees can be found via LinkedIn Pro, which proved to be a useful tool to locate staff within a given area of the organizations. Once staff were identified and contacted, a subsequent challenge related to the difficulty of finding informants willing to share their observations, beyond the designated spokesperson within either organization. In practice, I would continually be referred to a selected point of contact when I tried to contact staff with different expertise across the organization. Although I explained that gathering different perspectives across the organization was a point in itself, staff would insist that the identified spokesperson would be the best person to answer my questions. Eventually, I almost gave up on interviewing people and considered basing the analysis solely on publicly available statements such as the Zuckerberg archive,[1] combined with the relatively large amount of publicly available presentations, interviews, and so on from Google and Facebook founders and policy staff. Instead, I decided to revise my data collection strategy and focus on spaces where the two companies were present, such as Internet policy meetings (e.g., the Internet Governance Forum, EU meetings related to Internet governance) and more specific industry or research events (e.g., Global Network Initiative Public Learning Forum, Google Developer Groups meetings). This proved to be a good strategy, both as a means of establishing follow-up contact and in terms of engaging in discussion outside the more formal interview setting. I realized that when I made personal contact with Google and Facebook staff—for example, at a public meeting—they would often be more willing to be interviewed subsequently. Another useful strategy was to talk to local organizations that had some level of cooperation with the companies and were willing to introduce me to specific staff members. In sum, gaining access to interview subjects within the companies required an extensive amount of work compared with previous data collection I have been involved in—for example, among activists, civil society organizations, and government officials.

The Interview Situation

Interviewing elites posed methodological challenges to the empirical data collection. In addition to general interview considerations such as assessing the appropriate length of an interview, gaining the trust of respondents, dealing with respondents not answering questions, and closed- versus open-ended questions (Kvale 1997), specific issues relate to interviewing elites. The literature covering these methodological challenges has grown (Dexter 1970; Harvey 2011; Mikecz 2012). The methodological issues in elite interviewing involve issues of both validity (How appropriate is the interview format to the task at hand? Will the interviewer gain valid data?) and reliability (How consistent are the results?) (Berry 2002). Elites are often more likely to attempt to control the interview and be more particular about the questions they are willing to answer than other interview subjects (Harvey 2011, 439). This prompts researchers to consider how they present themselves and to show that they have done their homework because often elites might consciously or unconsciously challenge them on their subject and its relevance (Zuckerman 1972).

I conducted approximately 20 interviews with staff from Google and Facebook as part of my data collection. The interviews were via face-to-face meetings or done remotely via Blue Jeans (Facebook) or Google Hangouts (Google). I also visited the US and international (Dublin) headquarters of both companies, both for interviews and to attend meetings. Most of the respondents were staff members with responsibility for public policy, privacy, community operations (Facebook), and removal requests (Google).

However, I also conducted interviews with technical staff (Google) and staff working on education and user experience (Google). Two aspects of the interviews proved especially challenging: The first was how to handle an interview in which the topic is sensitive, the interview situation is restricted, and the respondents are cautious on what type of information they convey. In the second, challenges related to teasing out the implicit sensemaking and taken-for-granted context that effectively shape the framing and governance of human rights, such as the right to privacy and freedom of expression, within the two companies.

In relation to the first set of challenges—the constrained interview—I tried to leverage my own expertise in the field and to create an atmosphere of peer-based conversation rather than interview per se. Because of my previous work on Internet rights and freedoms (Jørgensen 2013) and because access to respondents had proved so difficult to obtain, I was very aware that the topic of my research was politically sensitive within both organizations. As an illustration of this, none of the respondents would allow me to record the interview, which meant that I had to rely on extensive note-taking during the conversations. Writing notes during the interview had the disadvantage of drawing attention away from the interview, limited eye contact between the interviewer (me) and the interviewee, and at times made it difficult to ensure that the interview stayed on track. However, note-taking also has some advantages compared with recording. While typed notes provide a weaker description of the interview, they potentially provide more detailed off-the-record information (Byron 1993). In other words, while recording provides a detailed record of the interview, it may result in less information being conveyed because it limits the respondents' willingness to talk. In contrast to previous research that has found note-taking to encourage more frank and detailed responses (Dexter 2006), I did not find the responses to be significantly richer than recorded answers to similar topics I have found in the public domain (e.g., recorded panel discussions or question and answer sessions). I did find, however, that note-taking during the interview forced me to focus on key messages and to document those messages instantly.

To facilitate a trustful and open atmosphere, I would start each interview by explicitly addressing an existing controversy—between, for example, privacy advocates and Facebook—and outline how my research hoped to engage with these existing debates. My intent was to demonstrate solid knowledge on the topic and a contextual and political understanding of the issues. In many cases, this would generate a more open and friendly tone; however, my general sense was that the interviewees were extremely guarded and synchronized in their responses, and it proved difficult to get respondents to elaborate more broadly on company practices and policy development. While research indicates that the interviewer can use silences to create a tension that may lead to more detailed answers (Berry 2002), I was generally trying to keep the conversation going. Given the sensitivity of the topics, I was afraid long silences would contribute to an awkward atmosphere and thus make the respondents less likely to elaborate on their answers.

Elites generally dislike closed-ended questions, which confine them to a restricted set of answers: "Elites especially—but other highly educated people as well—do not like being put in the straightjacket of close-ended questions. They prefer to articulate their views, explaining why they think what they think" (Aberbach and Rockman 2002, 64). While I took note of this in my interview style, I was also aware that I often had limited time to speak with the respondents and therefore a structured approach was necessary to obtain a focused response in a short time frame. Consequently, I conducted most interviews in a semistructured form using interview guides that reflected the key themes of my research. The semistructured format allowed improvisation and following subjects that appeared during the interview, including issues that were not part of the interview guide. Similarly, some areas were relevant to some respondents but not to others; hence, the thematic structure I followed varied. Despite the cautious approach toward staff interviews that both companies had demonstrated, some respondents were surprisingly generous with their time and would allow up to 90 minutes for an interview. In general, the face-to-face interviews lasted considerably longer than the interviews conducted remotely.

Although I experimented with relatively open-ended questions, the answers were in most cases fairly short and seemed standardized. However, there were also respondents who did not follow this pattern and would go into longer elaborations on a topic. In some cases, the respondents would advertently or inadvertently not answer a question. In these circumstances, I would rephrase and ask again. If they still did not answer the question, I would continue my line of questions and then circle back to the original question later in the interview. While this circling-back approach proved useful in some interviews, that was more the exception than the rule. Arguably, staff working with public policy within either organization are accustomed to dealing with policy makers, media, civil society, and so on and are very conscious of how they respond to politically sensitive questions.

When I compared the data collected via interviews with the publicly available presentations—for example, on content moderation practices—I found that the narratives and examples used were almost identical. In other words, after having listened to a few public talks on a given topic (e.g., transparency reporting) I could anticipate how a question related to that topic would probably be answered by one of my respondents. Irrespective of whether the topic concerned the core company values (e.g., the importance of free speech and privacy), the approach toward human rights (e.g., the description of corporate safeguards against government overreach), or governance challenges (e.g., content removal mechanisms), the responses were closely aligned across the people I talked to and even across the two companies for themes common to both organizations. In terms of corporate culture, it is interesting that storytelling is so uniform around what the company is doing and why it is doing it, in particular the role that technology is seen to play in making the world a better place (see later discussion). In terms of critical self-reflection, my experience was that when I referred to a given company policy that had been criticized (e.g., Facebook's Free Basics or Google's merger of its privacy policies), it was often presented as a factual misunderstanding, thus something that could be remedied by more information on what the company was actually doing. Few of the respondents were willing to reflect substantively on criticism—for example, from researchers or civil society organizations—of a specific practice.

The second set of challenges relate to identifying the implicit sensemaking and taken-for-granted context that inform the human right narratives within the two companies. Respondents seemed cautious and guarded when talking about corporate practices and human rights, yet it became clear to me after a couple of interviews that they shared an interpretative frame that governed their understanding of the relationship between the company and the framework of international human rights law. To tease out this frame, and to make it more explicit, I would first take note of the vocabulary and metaphors staff used to talk about specific human rights issues. I noted, for example, that the respondents repeatedly highlighted the liberating and transformative role of technology and thus of technology companies in making the world a better place. "We believe that, if more people are connected, the world will be more successful economically and socially" (#4, Facebook 2015). "What we do will help make the world a better place" (#3, Google 2015). Both the Facebook narrative (giving all individuals the ability to share and connect) and the Google narrative (making all the world's information accessible) seem deeply rooted in a belief system of doing good for society. Moreover, there is a strong narrative around human rights threats posed by either *repressive* or *controlling* (regulating) governments. "Our main focus is to minimize harms from governments" (#7, Google 2015). In terms of repressive governments, the respondents would refer to countries such as China, Russia, Iran, Turkey, and Egypt that are known for Internet censorship (Clark et al. 2017), whereas they would say controlling governments were those European countries seen as paternalistic with regard to data protection regulation.

To validate the interpretative frame I had drawn from the interviews, I would refer to it in follow-up questions or in the last part of the interview. I would ask, for example, if it was correct to note that governments are perceived as the primary threat to the protection of human rights online, and I would continue by asking about the institutional mechanisms established to counter governmental overreach. I fine-tuned my interpretation on the basis of the responses and added nuance to the description of the corporate storytelling. For example, my respondents all emphasized privacy as an important norm; however, I realized that the respondents referred to a very specific understanding of user privacy. I therefore went through several questions asking staff more explicitly what privacy protection at Google and Facebook entailed, and I found that it refers solely to *user empowerment* and *user control* while using the products and not to data minimization, which is core to data protection regulation and thus to online privacy protection in Europe. Whereas reference to data minimization would challenge the underlying logic of the online business model based on data maximization, the framing of privacy as user control may be accommodated through various privacy settings, as well as improved user information. In short, the extensive data collection that the data-driven business model relies on is an inherent element in the interpretative frame in both organizations. The business model functions as a taken-for-granted context and is seen as an integral and essential component in providing a "free" service (Jørgensen 2017).

Analysis

The third and final theme relates to the data analysis, specifically how to identify and describe the core human rights narratives and their reflection at the technical, legal, and governance levels. For theme analysis, I used a combination of meaning condensation and narrative structuring (Thagaard 2004, 158–163). I searched for theme-related points in the interview notes and tried to identify persistent explanations across the data collected.

The analysis revealed that public policy staff at Facebook and Google are aware of, and refer to, the international framework of human rights law and practice. When staff were elaborating on the role of the company vis-à-vis human rights, there was a very persistent narrative around state-caused threats to the free flow of information, invasive behavior from repressive states, skepticism toward state regulation of company practices (e.g., detailed data protection regulation), and a strong belief in self-regulatory measures. Although there are certainly cultural differences between the two companies, they both seem guided by a strong belief in the power of technology and in finding technical solutions to complex societal problems such as power inequality and uneven access to information. Generally, the interviewees shared a belief that their products contribute positively to social and economic development and thus to the enjoyment of human rights for everyone.

When trying to detect in more detail the nature of the human rights commitment expressed by staff at both companies, subthemes emerged. First, there was a strong emphasis on the state-centric nature of human rights violations, focusing almost entirely on user harms caused by states. While it is true that human rights law holds no direct obligations for private actors, this does not imply that private actors do not cause human rights harm. On the contrary, there is an increasing recognition of the potential negative effect on human rights from private actors and a call on companies to mitigate it (Knox 2012). In fact, several soft law standards provide guidance to companies on their responsibility to respect human rights.[2] When questioned about their human rights responsibility, some of the respondents referred to the UN Guiding Principles on Business and Human Rights, which represent the most recent soft law standard in this domain (UN Human Rights Council 2011). The guiding principles reaffirm states' obligation to ensure that businesses under their jurisdiction respect human rights but also outline the human rights responsibility of business, independent of how states implement their human rights obligations. Anchored in the UN "Protect, Respect, and Remedy" framework, the principles establish a widely accepted vocabulary for understanding the respective human rights roles and responsibilities of states and business. None of the respondents, however, mentioned that the UN Guiding Principles assert a global responsibility for businesses to avoid causing or contributing to adverse human rights impacts through their own activities and to address such impacts whenever they occur.

Second, the state-centric narrative was informed by several cases in which states had tried to assist technology companies in censorship, shutdowns, or surveillance. This is not surprising, as the ability of states to compel action by technology companies has been the source of some of the most widespread and severe human rights violations facilitated by the technology sector (Sullivan, 2016). In response, several of the major Internet companies cofounded the Global Network Initiative (GNI) in 2009. In the interviews, the GNI was mentioned time and again as a platform for multistakeholder engagement and as a sector-specific example of developing standards and guidance on how to respond to—and mitigate—overreach by states. The benchmarks and guidance produced by the GNI confirm the state-centric framing and thus is a narrative in which governments make more or less legitimate requests and companies respond and push back against such requests. With the increasing power that technology giants such as Google and Facebook hold in the online ecosystem and the standard setting provided by the UN Guiding Principles, this take on the human rights responsibility of companies is far too narrow. Arguably, there is a rising expectation from scholars and civil society organizations that the companies not only safeguard their users from overreach by states but also assess and mitigate negative human rights impact caused by their own business practices.

Third, safeguarding user content from state overreach was spoken of as a freedom of expression issue; however, the state-centric narrative effectively meant that the companies' enforcement of self-defined boundaries for allowed content ranging from clearly illegal to distasteful—and anything in between—was not identified as a human rights issue (Jørgensen 2018). Whereas state requests for takedowns rested on some level of transparency, terms of service enforcement appeared largely invisible and guarded from any kind of public scrutiny. Likewise, for privacy, employees at both

Facebook and Google emphasize that they pay great attention to privacy and push back against government requests for user data with due diligence standards. As an example of the commitment to privacy, both companies mention that an extensive system is in place to detect and react to privacy problems whenever a new product or product revision is introduced. In contrast, the underlying collection and use of personal data is not framed as a privacy issue. Privacy is repeatedly addressed as user control within the platform and not as limits to data collection per se. The idea of personal data as the economic asset of the Internet economy is presented as a given, and there is no recognition that the business model may, at a more fundamental level, preclude individual control. Thus, both companies praise the Internet's emancipatory potential and their role in facilitating this potential worldwide while not acknowledging that the infrastructure and services they control are key to people's ability (or inability) to exercise their rights. The strong belief in the liberating power of technology echoes the US online freedom narrative (Morozov 2011), which largely focuses on threats to the free and open Internet from repressive governments. It pays limited attention to the fact that rights such as freedom of expression, freedom of information, freedom of association, and so forth, are being exercised on private platforms—and thus the boundaries and modalities for these rights and freedoms are set by companies and outside the direct reach of human rights law. Moreover, that users' social interactions on the platforms are directly linked to revenue reinforces the imbalance between these companies and their users. Since I finalized the research, these issues have only increased on the public agenda, and data scandals such as that of Cambridge Analytica have prompted policy makers in Europe and the United States to discuss options for regulating the business practices of technology giants such as Google and Facebook. In 2018, the UN special rapporteur on freedom of expression dedicated an entire report to examining how states' and social media companies' regulation of user-generated content may affect freedom of expression (Kaye 2018). As part of the report, the rapporteur proposes a framework for content regulation based on human rights law.

Conclusion

In conclusion, I will examine how the human rights framework relates to and informs the broader field of Internet governance research.

First, the human rights framework is increasingly used by the companies that govern Internet infrastructure and services, not only major Internet platforms but also companies and corporations in the domain name system—as pointed to by Braman in chapter 2 and Musiani in chapter 4. Gatherings of the Internet Corporation for Assigned Names and Numbers (ICANN), for example, have had several sessions related to how ICANN understands and operationalizes its human rights commitment—at ICANN, registry, and registrar levels.[3] The increasing attention to the human rights framework in the technology sector is also demonstrated at the annual RightsCon summit,[4] which brings together companies, policy makers, and civil society groups from around the globe to discuss the intersection between technology use and regulation and the human rights implications of these practices.

Second, the increasing uptake of the human rights discourse in the technology sector contributes to defining public expectations and thus has policy implications. The narratives at Facebook and Google situate human rights issues firmly within a free market discourse in which companies provide products to users and are allied in protecting their users against repressive governments. This framing iterates the power of the free market and users' ability to choose between different products while not recognizing the negative impact that these platforms may have on users' rights. Essentially, it makes a difference whether Facebook is framed as an important part of the public sphere and public debate, with attached obligations, or whether it is seen as merely one product among many competing products in a free market. Likewise, for Google. Does it provide a social infrastructure whose governance has a direct impact on its users' ability to exercise rights and freedoms, or is it merely one among many search engines that users may or may not choose to use? The competing framings are important in understanding the conflicting expectations and their implications for the protection of rights within these spaces. While it is important that the technology sector address its human rights responsibility, its discourse and practice need to acknowledge that human rights impact assessment, mitigation, and due diligence must include all business processes and not only those involving state intervention.[5] The attention to the information and communication technologies sector by the UN special rapporteur on freedom of expression, David Kaye, is an important step in this direction (Kaye 2016). In short, we should be attentive to the risk of private companies capturing the human rights discourse without truly acknowledging that they themselves are powerful and potentially rights-violating actors in the ecosystem of human rights protection.

Third, the corporate uptake of the human rights discourse may divert attention away from the legal obligations of states to ensure that the rights of users are protected in the realm of private actors. In other words, states breach their human rights obligations both when abuse can be attributed to them and when they fail to take appropriate steps to prevent, investigate, punish, and redress private actors' abuse (UN Human Rights Council 2011). It is important to keep in mind that human rights law set out to protect the individual from abuse by power, and thus address a power imbalance, therefore there will be conflicting interests involved in human rights protection. A discourse that fails to acknowledge the conflict between, for example, users' right to privacy and a business model based on harvesting user data have not seriously dealt with the human rights impact of company practices. Likewise, if the discourse fails to recognize the potential conflict between standards for freedom of expression and terms of service enforcement, that effectively influences the boundaries for allowed speech for billions of users.

Finally, while there is a need to critically engage with private actors on their human rights impact and responsibility, one must be realistic about the enormous economic interests vested in a specific framing and thus about what can be achieved in terms of dialogue and voluntary commitment. In addition, it is important to recognize that while there are essentially different discourses on these topics inside and outside the companies, there are increasing cooperation and streams of funding from technology companies toward advocacy, policy, and research in this area. In this increasingly blurred landscape, it is important to distinguish and critically address both the legal obligations of states and the soft law responsibility of companies to assess and mitigate all business practices that may have a negative impact on human rights. As stressed by DeNardis in chapter 1, Internet governance covers "complex and multivariable points of control" that can be used to advance fundamental rights and freedoms or harm them. With this in mind, the blurred landscape of multistakeholderism and public-private corporations that surround the Internet governance domain poses specific challenges for conducting Internet governance research. It is thus crucially important that the researcher can interrogate, navigate, and study her field while retaining full integrity as a researcher. While this is a well-known challenge in many fields of research, it calls for hypercritical awareness in a field where the stakes are so high in terms of both economic and societal interests.

Notes

1. The Zuckerberg archive contains all public talks by Mark Zuckerberg in the period 2004 to 2019. Available at https://www.zuckerbergfiles.org/.

2. See *OECD Guidelines for Multinational Enterprises*, available at http://mneguidelines .oecd.org/text/, and *ILO Declaration on Fundamental Principles and Rights at Work*, available at http://www.ilo.org/declaration/lang--en/index.htm.

3. See, for example, *WS2—Enhancing ICANN Accountability*, available at https:// community.icann.org/display/WEIA/Human+Rights, and *NCSG-ICANN and Human Rights*, available at https://meetings.icann.org/en/marrakech55/schedule/mon-ncsg -human-rights.

4. See the RightsCon Costa Rica 2020 website, https://www.rightscon.org/.

5. See Götzmann 2019 for a comprehensive overview of the state-of-the-art of human rights impact assessment in the global business context, including sector-specific case studies.

References

Aberbach, J., & Rockman, B. (2002). Conducting and coding elite interviews. *PS: Political Science and Politics, 35*(4), 673–676.

Ananny, M., & Gillespie, T. (2016, October 22–23). *Public platforms: Beyond the cycle of shocks and exceptions*. Paper presented at the meeting of Internet, Policy and Politics conference. Oxford Internet Institute, Oxford, UK.

Barnett, M. N., & Finnemore, M. (2004). *Rules for the world: International organizations in global politics*. Ithaca, NY: Cornell University Press.

Berry, J. (2002). Validity and reliability in elite interviewing. *PS: Political Science and Politics, 35*(4), 679–682.

Byron, M. (1993). Using audio-visual aids in geography research: Questions of access and responsibility. *Area, 25*(4), 379–385.

Calhoun, C. (1995). *Critical social theory: Culture, history, and the challenge of difference*. Cambridge, MA: Wiley-Blackwell.

Clark, J., Faris, R., Morrison-Westphal, R., Noman, H., Tilton, C., & Zittrain, J. (2017). *The shifting landscape of global Internet censorship*. Berkman Klein Center for

Internet & Society Research Publications. Retrieved from http://nrs.harvard.edu/urn
-3:HUL.InstRepos:33084425

Dexter, L. A. (1970). *Elite and specialized interviewing*. Evanston, IL: Northwestern
University Press.

Dexter, L. A. (2006). *Elite and specialized interviewing*. Colchester, UK: European Consortium for Political Research.

Edwards, P. N. (1997). *The closed world: Computers and the politics of discourse in Cold
War America*. Cambridge, MA: MIT Press.

Flichy, P. (2007). *The Internet imaginaire*. Cambridge, MA: MIT Press.

Götzmann, N. (Ed.). (2019). *Handbook on human rights impact assessment*. Cheltnen-
ham, UK: Edward Elgar.

Harvey, W. S. (2011). Strategies for conducting elite interviews. *Qualitative Research,
11*(4), 431–441.

Huberman, A. M., & Miles, M. B. (1994). Data management and analysis methods.
In N. K. Denzin & Y. S. Lincoln (Eds.), *Handbook of qualitative research* (pp. 428–444).
Thousand Oaks, CA: Sage Publications.

Jørgensen, R. F. (2013). *Framing the net: The Internet and human rights*. Cheltenham,
UK: Edward Elgar.

Jørgensen, R. F. (2017). What platforms mean when they talk about human rights.
Policy & Internet, 9(3), 280–296.

Jørgensen, R. F. (2018). Framing human rights: Exploring storytelling within Internet companies. *Information, Communication & Society, 21*(3), 340–355.

Jørgensen, R. F., & Desai, T. (2017). Privacy meets online platforms: Exploring privacy complaints against Facebook and Google. *Nordic Journal of Human Rights, 35*(2),
1–21.

Kaye, D. (2016, May 11). *Report of the special rapporteur on the promotion and protection
of the right to freedom of opinion and expression*. Geneva, Switzerland: Human Rights
Council.

Kaye, D. (2018, April 6). *Report of the special rapporteur on the promotion and protection
of the right to freedom of opinion and expression*. Geneva, Switzerland: Human Rights
Council.

Knox, J. H. (2012). The Ruggie rules: Applying human rights law to corporations. In
R. Mares (Ed.), *The UN guiding principles on business and human rights: Foundations and
implementation* (pp. 51–83). Leiden, Netherlands: Martinus Nijhoff.

Kvale, S. (1997). *Interview: An introduction to the qualitative research interview*. Copenhagen, Denmark: Hans Reitzels Forlag.

Long, N. (2001). *Development sociology: Actor perspectives*. London, UK: Routledge.

Mansell, R. (2013). *Imagining the Internet: Communication, innovation, and governance*. Oxford, UK: Oxford University Press.

Mikecz, R. (2012). Interviewing elites: Addressing methodological issues. *Qualitative Inquiry, 18*(6), 482–493.

Morozov, E. (2011). *The net delusion: The dark side of Internet freedom*. New York, NY: PublicAffairs.

Sullivan, D. (2016, June). *Business and digital rights: Taking stock of the UN guiding principles for business and human rights*. Association for Progressive Communications Issue Papers. Retrieved from https://www.apc.org/sites/default/files/BusinessAndDigitalRights_full _report.pdf

UN Human Rights Council. (2011, March 21). Guiding principles on business and human rights: Implementing the United Nations "protect, respect and remedy" framework. Report of the special representative John Ruggie. New York, NY: Author.

Williams, R., & Edge, D. (1996). The social shaping of technology. *Research Policy, 25*(6), 865–899.

Zuckerman, H. (1972). Interviewing an ultra-elite. *The Public Opinion Quarterly, 36*(2), 159–175.

9   Big Data Analytics and Text Mining in Internet Governance Research: Computational Analysis of Transcripts from 12 Years of the Internet Governance Forum

Derrick L. Cogburn

Since the late 1990s, Internet governance has been a critical topic for multidisciplinary academic research (Bygrave and Bing 2009; Cogburn 2003, 2005; Cogburn et al. 2005; DeNardis 2009; Goldsmith 2007; Mueller 2009, 2010; Paré 2003; Thierer and Crews 2003). One simple measure of the increasing interest in Internet governance research is Google Trends statistics. In figure 9.1 we see November 2005 as the height of popularity for web searches of "Internet governance." This was the final month leading up to the second phase of the World Summit on the Information Society (WSIS), held November 16–18, 2005, in Tunis, after which the United Nations Internet Governance Forum (IGF) was launched (Cogburn 2017). Internet governance was certainly important before WSIS, but this global meeting helped accelerate a broad, multistakeholder focus on the issues beyond the narrow technical and academic focus that had dominated the study of Internet governance since the mid-1990s (Cogburn 2017). In chapter 3, Mueller and Badiei highlight these same trends.

In some ways research on the narrower Internet governance domain of cybersecurity has eclipsed the broader study of Internet governance. During the same time period, 2004–2017, there was a steady rise in searches for the term "cybersecurity," with a relatively sharp increase starting in 2013 (figure 9.2). Many researchers turned to a focus on better understanding the political and strategic implications of decisions made by infrastructure providers (Musiani et al. 2016). This rise is correlated with a steady increase in actual cybersecurity attacks, including the March 2007 hack of TJX, parent company of T. J. Maxx (Pepitone 2014), the June 2010 Stuxnet attack against Iran's nuclear centrifuges (Kushner 2013), the November 2013 Target (Kassner 2015), September 2014 Home Depot (Vinton 2014), and the April and June

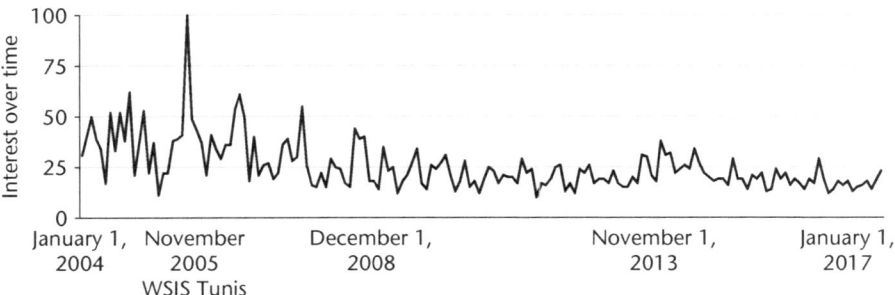

Figure 9.1
Google Trends searches for "Internet governance," 2004–2017.

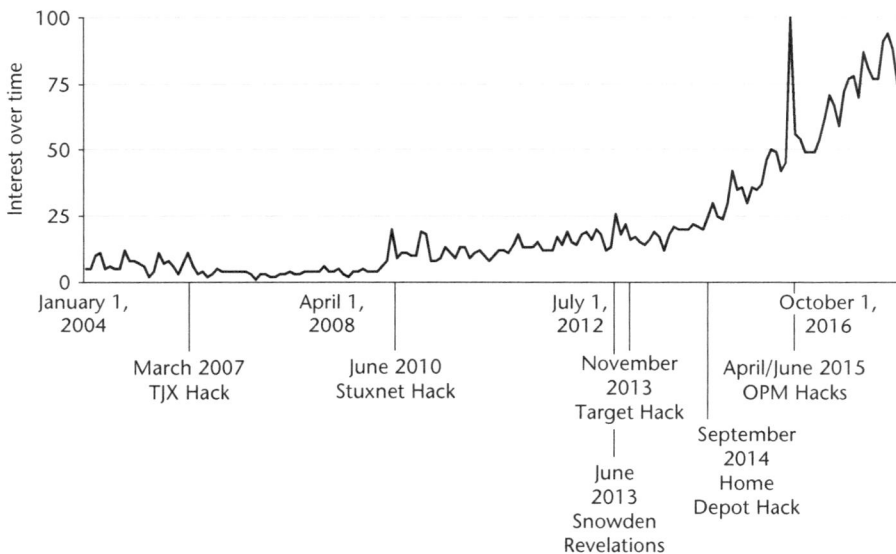

Figure 9.2
Google Trends searches for "cybersecurity," 2004–2017.

2015 Office of Personnel Management data breaches (Koerner 2016). This rise is also correlated with the revelations by Edward Snowden of US government widespread surveillance of the Internet, which began in June 2013 (Green-wald 2013). It also corresponded with the launch of the 2013 Barack Obama administration executive order on cybersecurity (Executive Order No. 13,636, 2013) and the subsequent National Institute of Standards and Technology (NIST 2014) Framework for Improving Critical Infrastructure Cybersecurity.

During the same period, the terms "big data" and "analytics" also became much more widespread. Between 2004 and 2012, Google Trends indicates search for the term "big data analytics" was relatively flat, with an average popularity score of 5. But then it exploded, jumping to 15 in January 2012 and to a high of 100 in March 2017. Figure 9.3 illustrates this trend. Schneider (2016) highlights some of the potential reasons for these trends, such as the increasing use of technologies that produce digital data, including unstructured text, and the corresponding increase in computational power available to analyze them.

Some describe big data, the Gartner Group being the first, by its 3Vs, or the volume, velocity, and variety of data that are available today (Laney 2001). Some scholars add veracity, variability, and value to those for understanding the concept of big data.

A particularly interesting type of data is underutilized: unstructured textual data. By unstructured, we mean text that has not yet been tagged, coded, or organized in some predetermined data model. Schneider (2016) estimates that up to 80 percent of the world's available data is unstructured text. The growing digital production and digitization of text adds significant amounts of textual data for analysis. This includes text on websites and in blog posts, speeches, meeting transcripts, email archives, reports, published articles, and especially social media (e.g., Twitter, Facebook, RSS feeds). Twitter is a particularly interesting source of unstructured textual data, making "nearly all of its data available via APIs [application programming interfaces] that enables [*sic*] realtime programmatic access to its massive seven-year archive" (Leetaru et al. 2013), and "Twitter users produce

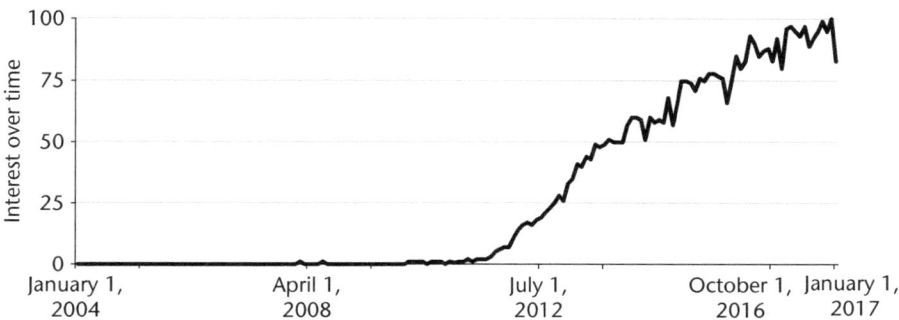

Figure 9.3
Google Trends searches for "big data analytics," 2004–2017.

8 billion words—every single day" (Kaisler et al. 2014). Of course, each of these data sources has "structure," but it is considered unstructured data because it has not yet been organized into a predetermined data model (e.g., into a spreadsheet with columns and rows).

In addition, each of these genres of data has its own characteristics that can be harnessed to augment analysis (Lee and Myaeng 2002). However, these many text-based datasets are not very large in terms of file size. One gigabyte of storage, an extremely small and common amount carried around by most students and faculty these days, can contain over 894,784 pages of plain text. A terabyte of data can contain 916,259,689 pages of plain text. It would take most humans an inordinately long time to read that many pages, but the data could be easily carried around on the average external drive. The big data of most social scientists or economists, however, especially those exploiting text as their data source, are not at the scale of the big data of earthquake engineering and upper atmospheric research. Thus, the "big" in "big data" is a relative term, with its meaning changing according to domain and the infrastructure available to the researcher. There is no singularly accepted definition of big data.

This increasing availability of data is coupled with a corresponding increase in computational tools, storage, and big data analytics and text mining software, including both commercial and open source options. The combination of this infrastructure and the available data allows us to combine insights from both quantitative and qualitative big data, but especially from unstructured textual data sources.

Purpose

This chapter, first, examines the substantive issues related to one of the major global institutional venues for debating issues related to Internet governance, specifically the annual UN IGF. Thus, this chapter focuses on identifying core themes and key issues discussed over the 12-year history of the IGF, and understanding which issues have remained constant, which have changed, and when they emerged or changed. Given the increasing importance of cybersecurity research in Internet governance, this chapter also explores the extent to which cybersecurity issues were debated at the IGF and the relationship between the NIST cybersecurity framework introduced by the Obama administration and related IGF debates. Second, the chapter demonstrates the potential of big data analytics and text mining techniques in Internet governance research.

These inductive and deductive text mining techniques are powerful tools to exploit the voluminous textual data available to Internet governance researchers and are also extremely useful for current research interests related to computational propaganda, or "fake news," and state- and nonstate-actor influences on national electoral systems through social media and other communication platforms. In chapter 7, Jardine highlights the continued growth of the Internet, including new users and content they create. He also argues for the suitability of quantitative methods for subsections of Internet governance research and particularly for cybersecurity research. The text mining approaches discussed in this chapter provide a nice complement to Jardine's arguments, by taking a quantitative analytical approach to analyzing text-based data, which are inherently qualitative. My use of a categorization model (dictionary) to measure the extent to which the NIST cybersecurity framework was present in the IGF debates reinforces his claims. In addition, Mueller and Badiei in chapter 3 discuss the widespread availability of materials on specific Internet governance institutions, such as the IGF, Internet Engineering Task Force, Internet Corporation for Assigned Names and Numbers, and others. There are many, many more, including the email archives, websites, and policy papers of transnational nongovernmental organization networks and of groups such as the Internet Assigned Numbers Authority Transition Committee. This chapter illuminates the reasons for analyzing those materials and making much better use of the voluminous resources available from these institutions for Internet governance research.

Conceptual Framework: Approaches to Text Mining

Text can contain substantial meaning and value to researchers. There are two important dimensions to text: semantics and syntax. *Semantics* refers to the meaning of words within their surrounding framework. *Syntax* is the structure of language, how individual words are arranged to make well-formed sentences and paragraphs. For decades, qualitative researchers have analyzed texts, doing deep and careful reading of relevant documents. As these qualitative research projects grew in size and complexity, computer-assisted qualitative data analysis software (CAQDAS) was developed to help facilitate this process. While extremely helpful, these CAQDAS tools still require researchers to closely read all documents and add codes to the text, developed a priori or in vivo while reading the documents.

The field of text mining is highly interdisciplinary and encompasses multiple theoretical approaches and methods with one common element—unstructured text as input information. Text mining has been aided by the widespread availability of machine-readable text. However, advances in the field of text mining, aided by concurrent increases in computational power and storage, have now accelerated the potential to use these techniques across a range of fields (Schneider 2016). With these tools, researchers can take unstructured text and transform it into a structured numerical format, based on term frequencies, and subsequently apply standard data mining techniques, finally unlocking the vast amount of valuable information in texts.

Many techniques are available to exploit the power and potential of big data analytics and text mining in specific research projects, including text classification, text clustering, ontology and taxonomy creation, document summarization, and latent corpus analysis. In general, there are two philosophical approaches to text mining, statistical and natural language processing. The statistical approach to text mining takes the "bag of words" route. It assumes there is value in the words themselves and does not require the analysts to understand the syntax of the words. In contrast, the natural language processing approach first tags parts of speech and then considers word and sentence structure. This study takes a statistical text mining approach, while recognizing the value of the natural language process approach.

Statistical text mining has two broad divisions—inductive and deductive, each with its own methodologies and techniques. The inductive approach asks broad exploratory questions about a large-scale text-based dataset, without specific a priori goals. For example, we can ask what key words and phrases characterize a dataset and determine what topics, themes, and trends exist. We can identify named entities within the dataset, including countries, people, organizations, and acronyms. Cross-tabulation determines how each element changes in relation to other key variables, such as date, region, and organizational type.

The deductive approach, in contrast, is confirmatory, allowing us to ask specific research questions of the data and to even test hypotheses. We can build, adopt, or adapt dictionaries (categorization models) to help us explore specific concepts in the dataset and to determine the degree of their presence or absence (Bengston and Xu 1995; Deng, Hine, and Sur 2017; Rousu et al. 2005). Variants of these models allow sentiment analysis, to characterize positive and negative sentiment, or polarity, within the dataset (Liu and Zhang 2012). Further, we can use supervised machine learning to develop classification models that allow us to predict text with a high degree of accuracy (Rousu et al. 2005). Through the use of these inductive and deductive techniques, we can begin to illustrate the tremendous potential of computational text mining for Internet governance and cybersecurity research.

Case Study: UN IGF

The WSIS action lines adopted at the end of the 2003 WSIS included promoting the continued development of the Internet with its potential impact on all aspects of the world (Cogburn 2017; International Telecommunication Union, n.d.). The action lines included four key references to Internet and Internet governance, and the 2005 WSIS Tunis Agenda mentions the Internet 80 times and Internet governance 30 times. At the conclusion of WSIS Tunis, participants adopted the Tunis Agenda, which in addition to coordinated implementation of the WSIS action lines, included a commitment to establish and support the UN IGF. The IGF was given an initial 5-year mandate and was subsequently approved for another 10 years.

The first IGF was held in Athens, Greece, in 2006, immediately after the conclusion of WSIS 2005 in Tunisia (Bygrave and Bing 2009). As of December 2017, there have been 12 IGFs, the last in Geneva. Going back to the first IGF in 2006, transcripts of sessions have been made available to the public on the IGF website (http://intgovforum.org), and over time, this process has increased and become more comprehensive. For example, in 2006, only 11 transcripts were made available, while in 2017, as many as 215 transcripts were made available (table 9.1).

Table 9.1
Number of IGF transcripts by year.

2006	2007	2008	2009	2010	2011	2012	2013	2014	2015	2016	2017	Total
11	14	14	15	114	61	8	63	138	162	205	215	1,020

Research Questions

To demonstrate the computational text mining techniques described above, this case study of IGF transcripts asks four broad research questions, two inductive and two deductive.

RQ1. What are the key themes, topics, and entities discussed at IGF over its lifetime?

RQ2. Which key issues have remained consistent at IGF, and which ones have changed?

RQ3. In what ways is the Internet of Things (IoT) represented at IGF?

RQ4. To what extent is the 2014 NIST cybersecurity framework represented at IGF?

Methodology

This study is organized using the cross-industry standard process for data mining (CRISP-DM) for text mining.[1] Since text mining is still a relatively new and somewhat unstandardized field, the CRISP-DM approach can provide a well-understood, documented, and somewhat standardized process for executing and managing complex text mining projects. The CRISP-DM for text mining has six stages through which each text mining project must proceed (figure 9.4). In Stage 1, the researcher determines the purpose of the text mining study: what the researcher wants to accomplish and the problem or opportunity identified by the researcher. In Stage 2 the researcher explores the availability and nature of the unstructured textual data to exploit. The researcher has to determine if the data are available, in what format they are stored, and in what quantity. Stage 3 focuses on preparing data, which could include data cleaning, preprocessing, applying stopwords (exclusion lists that remove common, insignificant words), and further data reduction techniques of stemming and lemmatization. In Stage 4, the researcher develops the models and specific techniques to analyze the data. In Stage 5 the researcher evaluates the results of the analysis. In Stage 6 the researcher deploys the results, in the form of recommendations and presentations. At any point along the way, the researcher may decide to go back to a previous stage or all the way to the beginning.

Using an automated commercial tool called SiteSucker,[2] the researcher collected all the publicly available IGF transcripts. This data collection

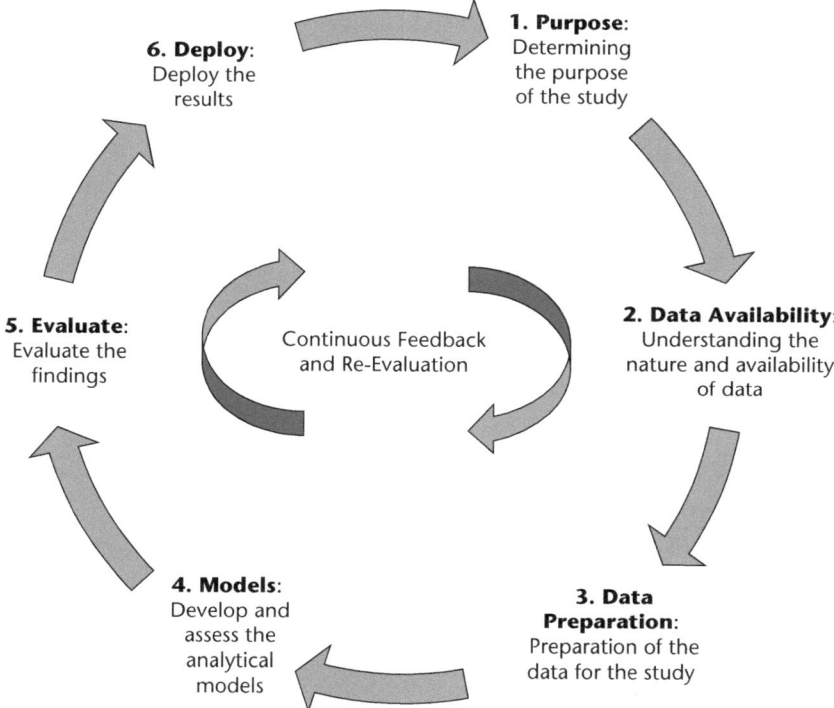

Figure 9.4
The CRISP-DM approach.

yielded 1,020 documents (made up of .txt, .html, .doc, and .pdf formats), in a file of 109.9 megabytes. The data collection represents the available transcripts of main sessions and workshops where available.

Once the data were collected, the researcher used a commercial software tool called Provalis ProSuite to organize the project and conduct the text mining.[3] There are open source software tools that can also perform this type of analysis, such as the R programming language and related packages. Most R programmers use RStudio, an open source integrated development environment (IDE), which makes it easy to install and use free and open source text mining packages such as tm, Rvest (for data collection), and tidytext. The first step is to build the corpus, which includes converting the textual data into numerical data, based on the word frequencies across documents. This corpus, containing all the available IGF transcripts, is constructed primarily as a document-term-matrix (DTM), which means each document is repre-sented as a row in the matrix, while each term (word) in each document is represented in the columns of the matrix, with a numerical value for how many times that term occurs in the document. Upon import, I used the file structure (organized by date of the IGF) to automatically create a Date vari-able for filtering the dataset by date and conducting a longitudinal analysis. I also used this Date variable for cross-tabulation analysis. Then, I preprocessed the data, applying a typical English-language stopword dictionary (or exclu-sion list) to remove frequent words that add little value to the analysis (such as "a," "and," "the"). The exclusion list may be modified for a specific dataset (e.g., words deemed important to include or remove in the analysis). I did not apply stemming or lemmatization, which preprocesses a textual dataset to reduce its overall size.

To answer the first two of the four research questions, I began with an inductive approach, a count-based evaluation, that focuses on term and document frequency, followed by phrase frequency. This is a common approach, and is one of the simplest techniques for text mining, similar to basic descriptive analysis of project variables in a statistical study. A word or phrase, an n-gram, occurring frequently in a dataset, with some impor-tant limitations discussed later, is considered important. In this analysis, I used the term frequency by inverse document frequency (TFxIDF) tech-nique: if a word appears frequently in a document, it is important; but if it appears in many documents, it is less important. This is a common text mining heuristic to identify important words and phrases in a corpus.

Next, I used an inductive technique called topic modeling, which essen-tially does exploratory factor analysis on the underlying numerical repre-sentation of the IGF transcripts to identify factors, which are interpreted as topics. However, unlike factor analysis, since the dataset is based on text, the software provides a textual suggestion of what the topic seems to repre-sent. I applied topic modeling on the entire dataset and separately for each of the 12 years. In addition, I identified key organizations, countries, acro-nyms, and people across the entire dataset, and again for each year, using a named-entity extraction tool.

To answer the third and fourth research questions, I took a deductive approach: hierarchical cluster analysis and categorization modeling (also known as dictionary development) to answer the third question on the IoT, and categorization modeling to determine the extent to which the 2014 NIST cybersecurity framework is included in IGF discussions for the fourth.

Hierarchical cluster analysis allows examination of the entire dataset for co-occurrences and to assess the themes or topics represented by specific clusters, such as a cluster that appears to represent the IoT.

Categorization modeling is an explicitly deductive technique. Essentially, it requires the researcher to develop a semantic representation of the concept of interest, and then apply that dictionary to the corpus to determine the extent to which that concept is present or absent in the text. Dictionary development generally starts with the broadest categories within the concept (for example, in sentiment analysis, these top categories tend to be the binary categories of positive and negative). Then, those broadest categories can be further divided into broad subcategories. Once the lowest levels of categories and subcategories have been determined, specific words, phrases, and rules (which allow you to formulate criteria for text that includes negations and specifications for proximity of words and phrases) can be developed. When any of these elements occur, they accrue to the subcategory, which in turn aggregates up to its category. This is a very powerful technique to identify the extent to which a specific concept the researcher is interested in exploring is either present or absent in the dataset. In this study, I developed a categorization model (figure 9.5) starting from the 2014 NIST cybersecurity framework core spreadsheet.[4] This framework has five primary categories (identify, protect, detect, respond,

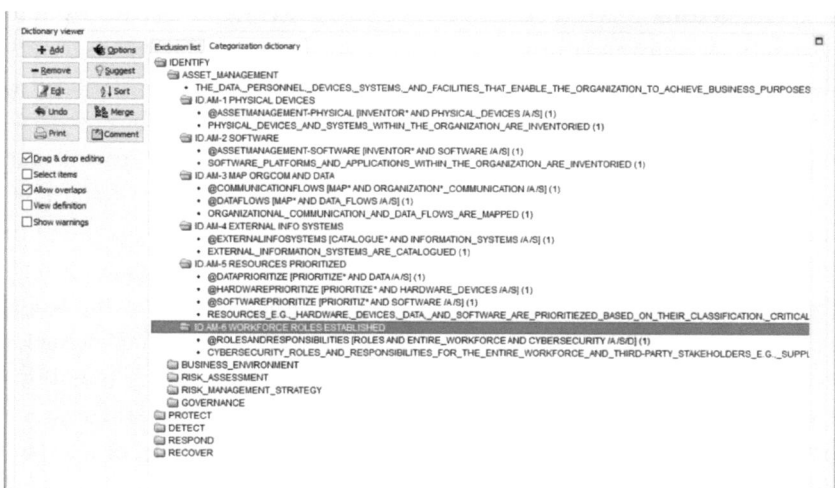

Figure 9.5
Overview of 2014 NIST cybersecurity categorization model.

and recover) and within each category are multiple subcategories and sub-subcategories. All these elements are captured in my categorization model. I deployed these categorization models across the entire 12-year period.

Similarly, I could have built another categorization model representing the EU Cybersecurity Framework and compared the degree to which each framework was represented in the dataset. Or I could have explored the dataset to assess the degree to which the priorities of one stakeholder—say, the private sector, represented by the group Business Action in Support of the Information Society (BASIS) and supported by the International Chamber of Commerce (ICC)—were represented in the dataset relative to, say, the statements of the civil society Internet Governance Caucus (IGC). I also could have used supervised machine learning to build a classifier to distinguish between the content of each stakeholder group and then deployed that classifier to assess which stakeholder group had the most influence in the IGF processes. It would be a little tricky to do this in the IGF context, because there are no concrete outcome documents of an IGF, but I used this technique to great effect in an analysis of stakeholder contributions to the NetMundial conference (Cogburn 2014).

Research Limitations and Challenges in Text Mining Internet Governance

As far as I know, the corpus of 1,020 IGF transcripts makes this the largest study to date of these important data, and these techniques have proved to be extremely valuable in studying Internet governance. However, this only scratches the surface, and this approach has important limitations. First, although the data for this study—transcripts from 12 years of the IGF—are tremendously revelatory, they do not cover all the workshops and side events associated with an IGF meeting, and they capture only formal statements from IGF sessions. This focus on what Goffman (1959) called "front stage" behavior ignores his argument of the importance "back stage" behavior can have on policy debates. Of course, much of the work of the IGF is accomplished outside the formal conference structure. Backstage behavior occurs during the coffee breaks, lunches, dinners, and the many receptions and parties associated with an IGF. Analyzing only the formal language is a major limitation of this approach. Also, there is the limitation of this particular dataset of who said what. It would be tremendously valuable to have each statement attributed to a particular actor, who could then be coded in a variety of ways (e.g., by multistakeholder sector or by regional, ethnic, or gender demographic). However, I would have had to be present for each session and have conducted more traditional participant-observation research or interviews or focus groups. Finally, most text analytics techniques wrestle with understanding syntactic meaning. Sarcasm, euphemism, and double entendre, all common in human language, continue to elude many of these computational approaches. Nonetheless, while limitations of these text mining techniques have no silver bullet, this study demonstrates some of their tremendous value in Internet governance research.

Findings

To answer the first research question, What are the key themes, topics, and entities discussed at IGF over its lifetime?, I applied TFxIDF to explore, first, keywords and then phrases across all 12 years. Figure 9.6 shows the top ten themes at IGF represented by keyword frequency, and figure 9.7 shows the

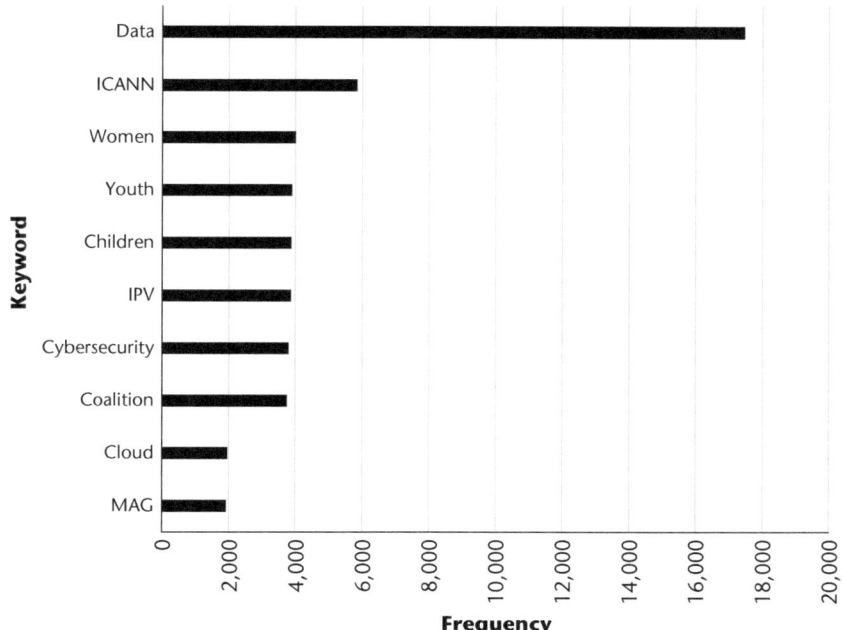

Figure 9.6
Key themes across 12 years of IGF represented by keywords.

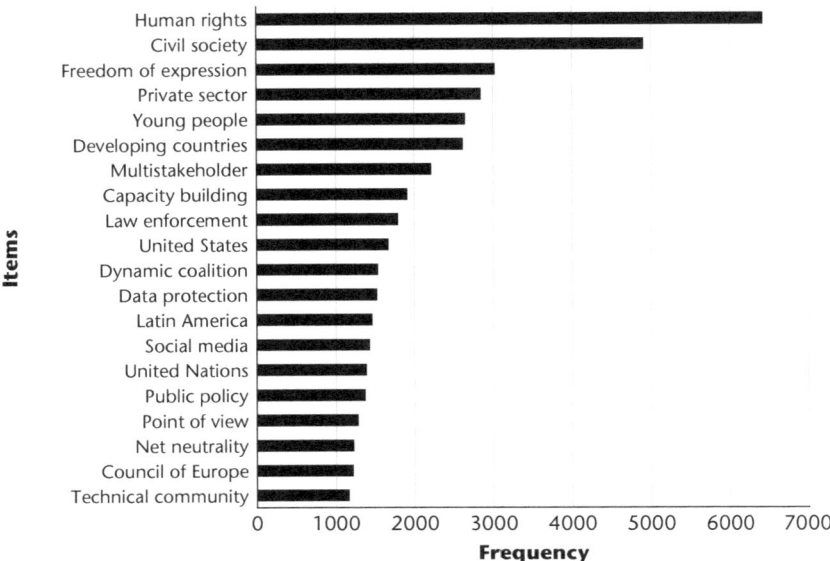

Figure 9.7
Key themes across 12 years of IGF represented by phrases.

top 20 themes of 12 years of IGF represented by phrase frequency. Then I used named-entity extraction techniques to identify all the people referenced in the dataset across 12 years. Figure 9.8 shows the top twenty names in the dataset.

To answer the second research question, Which key issues have remained consistent at IGF, and which ones have changed?, I explored the changes in key themes over the 12 years of IGF by identifying the top 20 themes at the beginning (2006; figure 9.9), middle (2011; figure 9.10), and most recently (2017; figure 9.11).

Also, when using the entity extraction tools, we identify the most frequently listed organizations, acronyms, countries, and people across all 12 years of the IGF. Figure 9.12 illustrates the top 25 organizations, acronyms, and countries across 12 years of the IGF. Figure 9.8 represents the top twenty names appearing in the IGF transcripts over the 12 years represented in this dataset.

To answer the third research question, In what ways is the Internet of Things (IoT) represented at IGF?, I conducted a cluster analysis across all 12 years. There were initially 60 clusters identified, representing significant

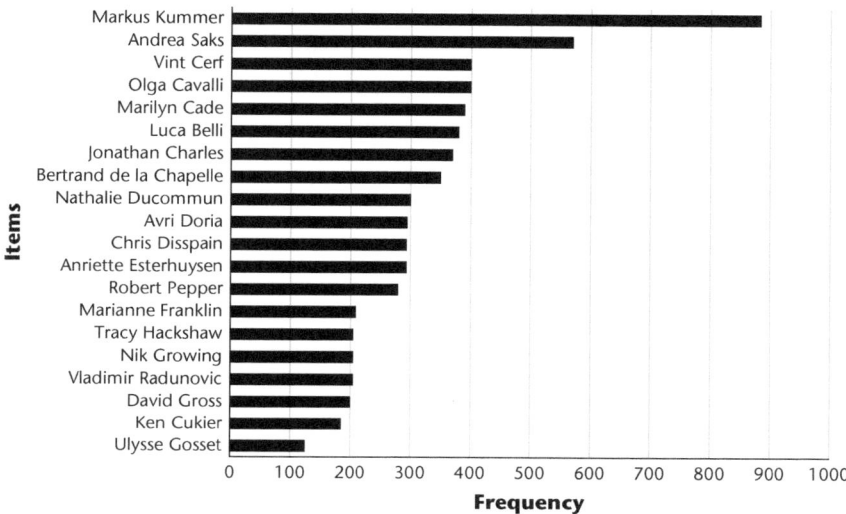

Figure 9.8
Top twenty person names at IGF across 12 years.

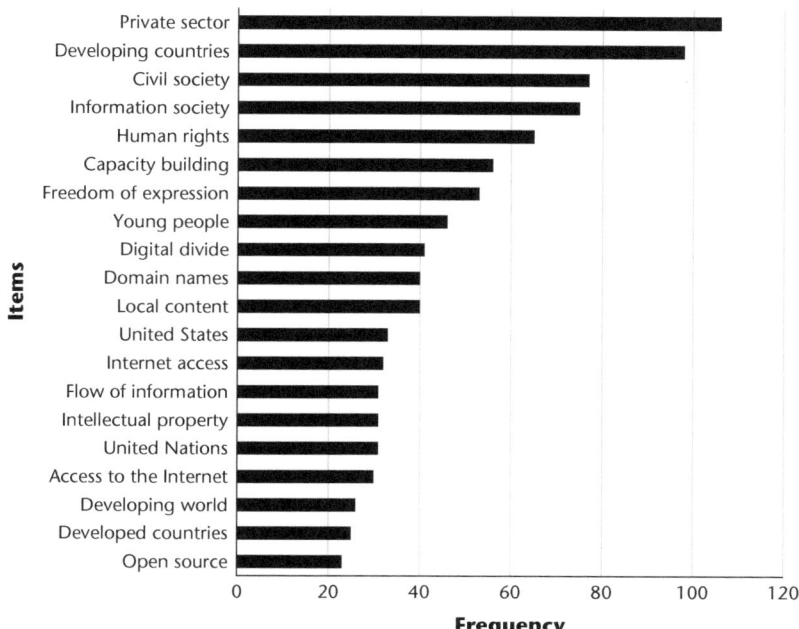

Figure 9.9
IGF top phrases 2006.

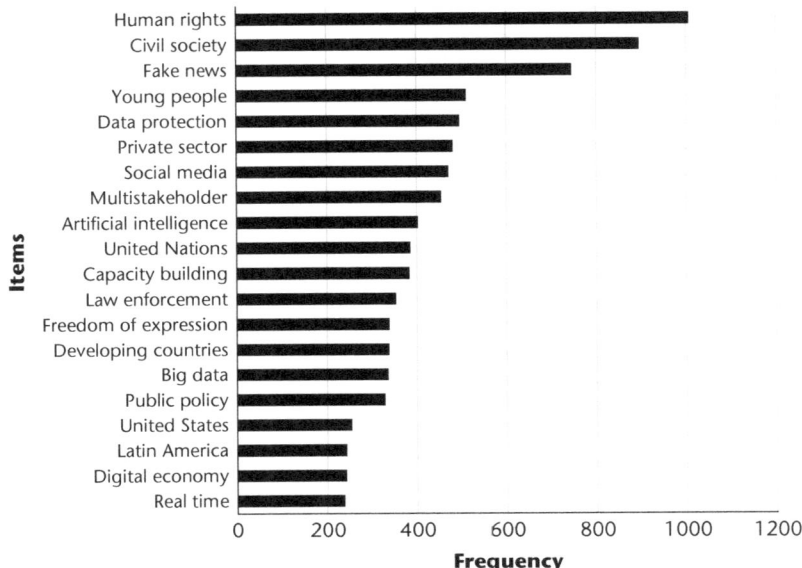

Figure 9.10
IGF top phrases 2011.

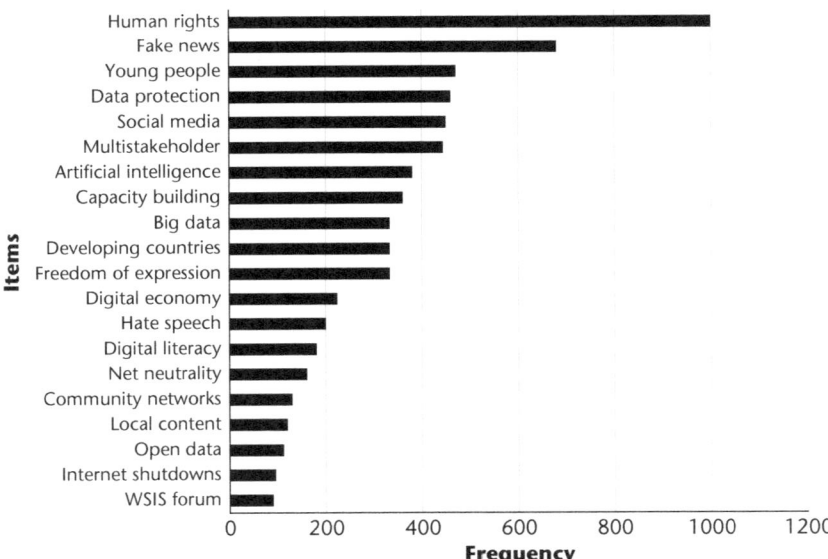

Figure 9.11
IGF top phrases 2017.

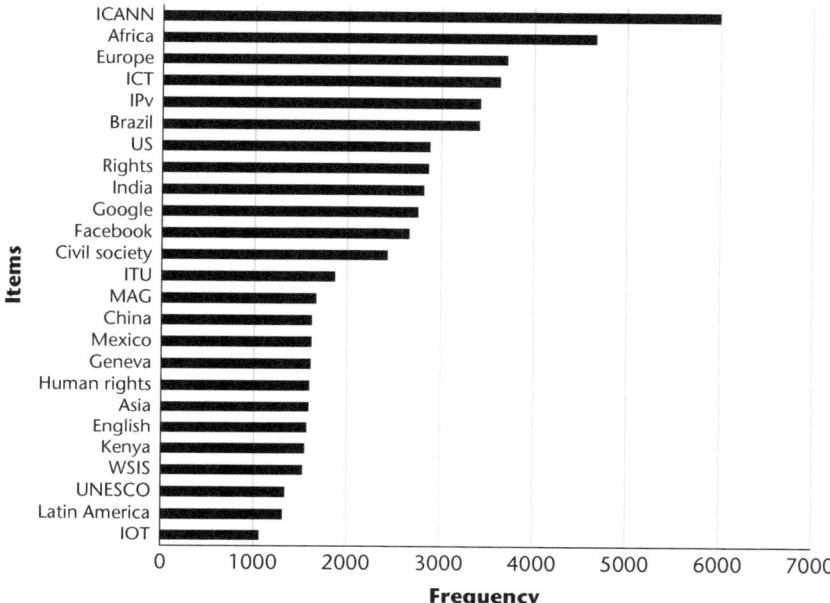

Figure 9.12
IGF entity extraction over 12 years.

thematic groupings. Figure 9.13 illustrates four of those clusters around child protection, capacity building, innovation in infrastructure (including broadband, mobile, net neutrality, and cloud computing), and smart cities and IoT.

Using the inductive technique of topic modeling to look across all 12 years of IGF and the middle and most recent IGFs in the dataset, I found that freedom of expression and human rights are the most durable and consistent topics. Internationalized domain names and mobile phones were taken over in the most recent IGF by fake news and media freedom and multistakeholder discussions. Table 9.2 highlights this topic modeling across the IGFs.

Finally, to answer the fourth research question, To what extent is the 2014 NIST cybersecurity framework represented at IGF?, I deployed the categorization model that captured all the primary categories, subcategories, and sub-subcategories of the framework.

These figures show the frequency of keywords later codified in the 2014 NIST cybersecurity framework when looking at the entire 12-year IGF

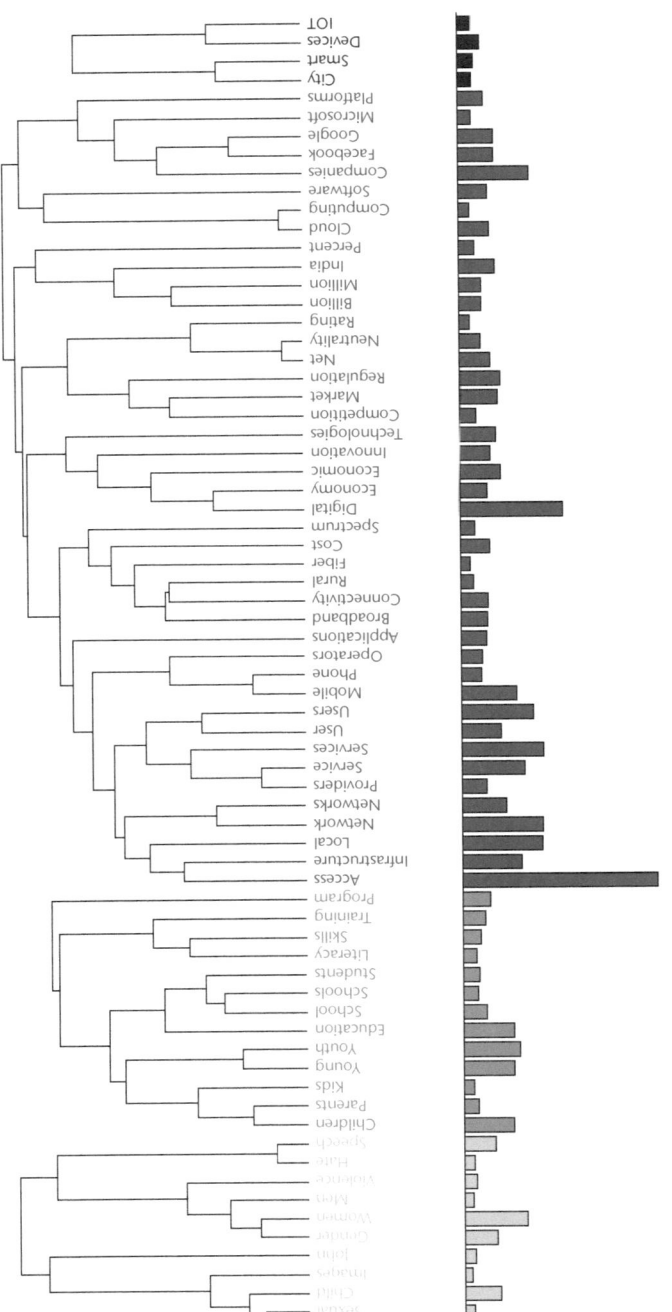

Figure 9.13
Partial illustration of the cluster analysis highlighting four of the 60 clusters.

Table 9.2
Topic modeling over 12 years of the IGF.

2008	2009	2010	2011	2012
IPV4-6 transition	DNS and IANA	Budapest convention	Human rights	Cybercrime
Child pornography	Accessibility	Mobile devices/ Wi-Fi	IDNs	IDNs
Enhanced cooperation		Human rights	Mobile phones	Disaster risk
Freedom of expression			Accessibility	

2013	2014	2015	2016	2017
Journalist/ bloggers	Human rights	Human rights	Human rights	IANA transition
Budapest convention	IANA transition	Net neutrality	Wi-Fi/fiber	Cybersecurity
Mobile devices	DNS	Child abuse	Children online	Fake news
Human rights	CERTs/CSIRTs	IANA transition	IXPs	Human rights
Intellectual property	Accessibility		SDGs	
IDNs			IANA transition	

dataset. Specifically, figures 9.15 and 9.16 allow us to drill down into the categorization model to the sub and sub-subcategories. For example, the most frequently occurring sub-category is PR.PT-3 Least Functionality. This subcategory corresponds with the overall main category labeled "protect," identified as "PR" in the categorization model. The protect "function" of the NIST cybersecurity framework has six categories, once of which is called "protective technology" (labeled PR-PT in the categorization model). The term "protective technology" refers to how an organization manages protective technology. As defined by NIST, in this subcategory "technical security solutions are managed to ensure the security and resilience of systems and assets, consistent with related policies, procedures, and agreements." Within protective technology, the third sub-subcategory (labeled PR-PT-3 in the categorization model) refers to the principle of least functionality. As defined by NIST, "the principle of least functionality is incorporated by configuring systems to provide only essential capabilities." In order to attempt to detect

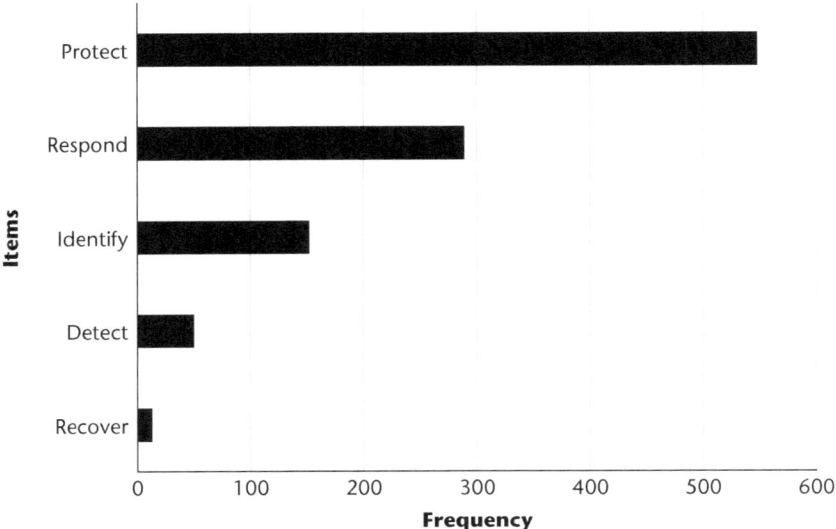

Figure 9.14
Distribution of NIST cybersecurity framework primary categories.

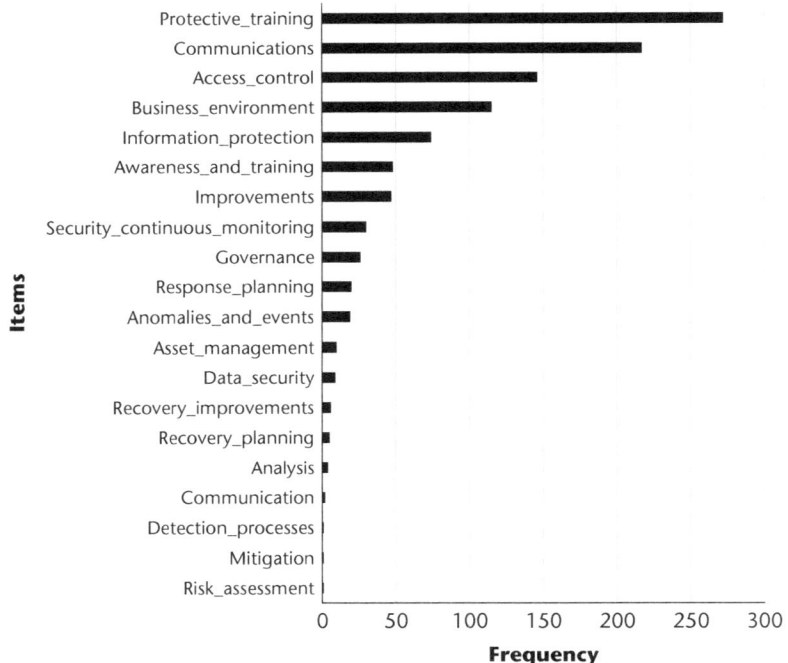

Figure 9.15
Distribution of NIST cybersecurity framework subcategories.

this level of the NIST cybersecurity framework, this component of the cybersecurity model included words and phrases related to the principle of least functionality. Figure 9.16 illustrates the finding that within the 12 years of IGF transcripts, the most frequently occurring component of the NIST cybersecurity framework (as represented by this categorization model) is the principle of least functionality. Each component and subcomponent of the NIST model is represented in this way, with RS.CO-3, the next most frequently occurring concept, representing the response function/category of the framework, and the communication subcategory. The response-communication subcategory focuses on response activities that are coordinated with internal and external stakeholders (e.g., external support from law enforcement agencies). The third element of the response-communication subcategory, labeled RS.CO-3, is designed to assess the degree to which "information is shared consistent with response plans." In this way, we can take this complex and highly detailed government framework and assess the degree to which each of its detailed parts is represented in a database of unstructured text.[5]

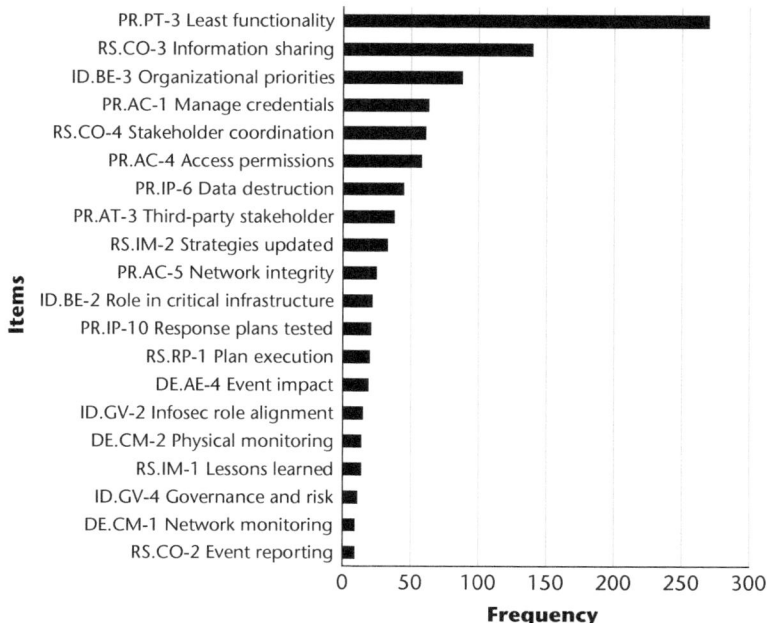

Figure 9.16
Distribution of NIST cybersecurity framework sub-subcategories.

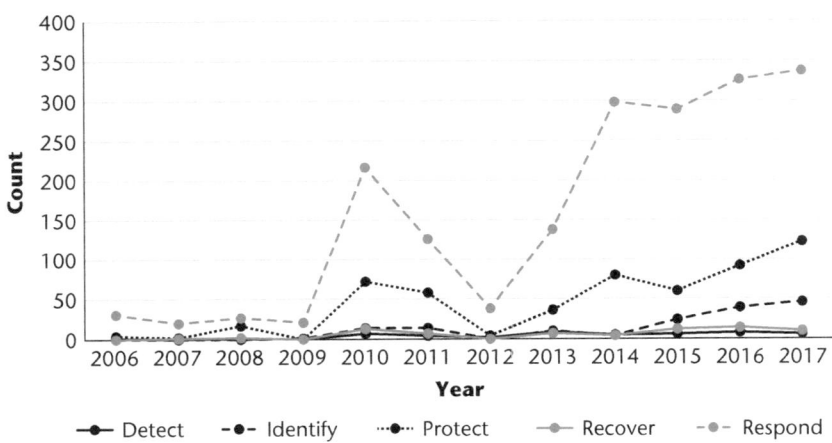

Figure 9.17
Year-by-year line chart distribution of NIST cybersecurity framework keywords.

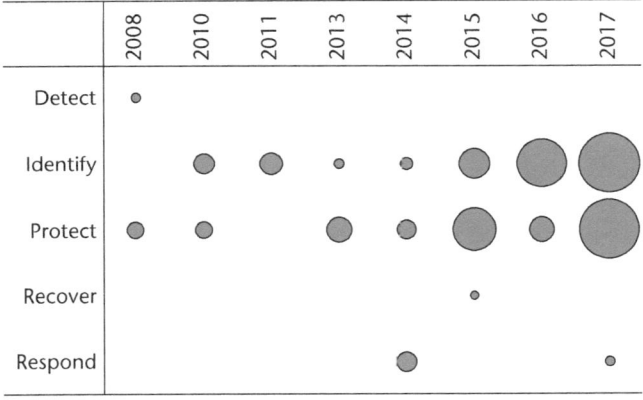

Figure 9.18
Year-by-year bubble chart distribution of NIST cybersecurity framework keywords.

Through the use of this detailed categorization model, we are able to estimate the degree to which concepts in the framework are present (or absent) in the 12 years of IGF transcripts. As such, we see that in 2014, the year the framework was introduced, there was a substantial increase in two of the five components: identify and protect. See figure 9.17.

Another way of viewing these data is via a bubble chart. In figure 9.18, we see that for these same two categories, identify and protect, of the NIST cybersecurity framework there was also an increase after 2014, when the first version of the framework was introduced.

Discussion

With this relatively brief analysis, I have identified the key thematic focus areas of the IGF over its 12-year lifespan. Even without the expensive and time-consuming participant observation most researchers studying the IGF would want to use, I identified key trends, patterns, changes in foci, and important actors—people—operating in this emblematic Internet governance institution. For example, one of the most surprising findings was how early and prominently disability and accessibility issues were included in the IGF. On the basis of topic modeling, we see accessibility appearing as a topic as early as 2009; it appears again in 2011 and 2014. Given the sensitivity of this analysis, and the difficulty of any term making it into the small list of core topics for any given year, this result is striking (only human rights [n=7] and the IANA transition [n=4] appeared more frequently). This finding is most likely a result of the work of the Dynamic Coalition on Accessibility and Disability and its long-term coordinator, Andrea Saks. In addition, while our named-entity extraction analysis identified the name of Markus Kummer appearing most frequently in the dataset, nearly twice as frequently as any other name (which makes sense, given that Markus was the head of the IGF secretariat, and a key leader of the movement within the UN), the second most frequently occurring name is Andrea Saks. Most of the remaining names on the list would be immediately recognizable to anyone studying or participating actively in the IGF over its lifetime.

I have also shown how, in line with Google Trends, cybersecurity topics have become increasingly important over the lifespan of the IGF, with a fairly clear correlation between the introduction of the NIST cybersecurity framework in the United States and these discussions globally at IGF. This finding helps to illustrate how a well-defined major power policy framework—in this case the US under President Obama—can potentially influence global discussions on that issue. My goal in this chapter was not to exhaust an analysis of the policy issues at IGF but to demonstrate the important role these big data analytics and text mining techniques can play in Internet governance research.

Conclusion and Future Research

This study identified some interesting substantive components of the IGF, including the key thematic focus areas over its 12-year lifespan and in a year-by-year comparison. In addition, although I have only scratched the surface, I believe I have demonstrated the power of big data analytics and text mining in Internet governance and cybersecurity research. In this and in other work I have tried to highlight the importance and potential impact of these techniques in monitoring and evaluating the UN's Sustainable Development Goals and implementation of the WSIS action lines.

Regarding future research, I have already highlighted some possibilities to pursue in the near term. Some of these will require adding more variables to the dataset, including type of gathering (e.g., main session, workshop), identifying which dynamic coalition organized the event, and finally, identifying the speaker by name or stakeholder grouping. However, nearer-term studies will focus on building other categorization models: first, to represent different approaches to cybersecurity in order to compare the degree to which each framework is represented in the dataset, and then to identify, represent, and compare other concepts, such as net neutrality and Internet freedom. I also believe exploring the dataset to assess the degree to which the priorities of various stakeholders are represented will be fruitful.

Notes

1. More information on the CRISP-DM process model (1999) is available at http://www.crisp-dm.org/.

2. See the SiteSucker website at https://ricks-apps.com/osx/sitesucker/index.htm.

3. Provalis ProSuite is available at http://provalisresearch.com/.

4. Framework V1.1 Core (Excel). NIST Framework for Improving Critical Infrastructure Cybersecurity. Available at https://www.nist.gov/cyberframework/framework.

5. For more detail, see the NIST cybersecurity framework website: https://www.nist.gov/cyberframework.

References

Bengston, D. N., & Xu, Z. (1995). Changing national forest values: A content analysis (Research paper NC-323). St. Paul, MN: US Department of Agriculture, Forest

Service, North Central Forest Experiment Station. Retrieved from http://www.nrs.fs
.fed.us/pubs/rp/rp_nc323.pdf

Bygrave, L. A., & Bing, J. (Eds.). (2009). *Internet governance: Infrastructure and institutions*. Oxford, UK: Oxford University Press on Demand.

Cogburn, D. L. (2003). Governing global information and communications policy: Emergent regime formation and the impact on Africa. *Telecommunications Policy, 27*(1–2), 135–153.

Cogburn, D. L. (2005). Partners or pawns? The impact of elite decision-making and epistemic communities in global information policy on developing countries and transnational civil society. *Knowledge, Technology & Policy, 18*(2), 52–82.

Cogburn, D. L. (2014, September 1). *Uncovering the conceptual antecedents of the NET-Mundial Outcome Document on the future of global Internet governance*. Paper presented at the Annual Symposium of the Global Internet Governance Academic Network (GigaNet), Istanbul, Turkey.

Cogburn, D. L. (2017). *Transnational advocacy networks in the information society: Partners or pawns?* New York, NY: Palgrave Macmillan Springer.

Cogburn, D. L., Mueller, M., McKnight, L., Klein, H., & Mathiason, J. (2005). The US role in global Internet governance. *IEEE Communications Magazine, 43*(12), 12–14.

DeNardis, L. (2009). *Protocol politics: The globalization of Internet governance*. Cambridge, MA: MIT Press.

Deng, Q., Hine, M., Ji, S., & Sur, S. (2017, January). Building an environmental sustainability dictionary for the IT industry. In *Proceedings of the 50th Hawaii International Conference on System Sciences*. Retrieved from http://hdl.handle.net/10125/41264

Exec. Order No. 13,636. (2013, February 12). *Improving critical infrastructure cybersecurity*. Retrieved from https://obamawhitehouse.archives.gov/the-press-office/2013/02
/12/executive-order-improving-critical-infrastructure-cybersecurity

Goffman, E. (1959). *The presentation of self in everyday life*. New York, NY: Doubleday.

Goldsmith, J. (2007). Who controls the Internet? Illusions of a borderless world. *Strategic Direction*.

Greenwald, G. (2013, June 6). NSA collecting phone records of millions of Verizon customers daily. *The Guardian*. Retrieved from https://www.theguardian.com/world
/2013/jun/06/nsa-phone-records-verizon-court-order

International Telecommunication Union. (n.d.). *WSIS Action Lines*. Available at https://www.itu.int/net/wsis/stocktaking/help-action-lines.html

Kaisler, S. H., Espinosa, J. A., Armour, F., & Money, W. H. (2014, January). Advanced analytics: Issues and challenges in a global environment. In *2014 47th Hawaii International Conference on System Sciences* (pp. 729–738). Waikoloa, HI: IEEE Computer Society.

Kassner, M. (2015, February 2). Anatomy of the Target data breach: Missed opportunities and lessons learned. *ZDnet*. Retrieved from https://www.zdnet.com/article/anatomy-of-the-target-data-breach-missed-opportunities-and-lessons-learned/

Koerner, B. I. (2016, October 23). Inside the cyberattack that shocked the US government. *Wired*. Retrieved from https://www.wired.com/2016/10/inside-cyberattack-shocked-us-government/

Kushner, D. (2013, February 26). The real story of Stuxnet. *IEEE Spectrum*. Retrieved from https://spectrum.ieee.org/telecom/security/the-real-story-of-stuxnet

Laney, D. (2001). 3D data management: Controlling data volume, velocity and variety. *META Group Research Note, 6*(70), 1. Retrieved from https://blogs.gartner.com/doug-laney/files/2012/01/ad949-3D-Data-Management-Controlling-Data-Volume-Velocity-and-Variety.pdf

Lee, Y. B., & Myaeng, S. H. (2002, August). Text genre classification with genre-revealing and subject-revealing features. In *Proceedings of the 25th annual international ACM SIGIR conference on Research and Development in Information Retrieval* (pp. 145–150). New York, NY: ACM.

Leetaru, K., Wang, S., Cao, G., Padmanabhan, A., & Shook, E. (2013, May). Mapping the global Twitter heartbeat: The geography of Twitter. *First Monday, 18*(5–6). Retrieved from https://firstmonday.org/article/view/4366/3654

Liu, B., & Zhang, L. (2012). A survey of opinion mining and sentiment analysis. In C. Aggarwal & C. Zhai (Eds.), *Mining text data* (pp. 415–463). Boston, MA: Springer.

Mueller, M. L. (2009). *Ruling the root: Internet governance and the taming of cyberspace*. Cambridge, MA: MIT Press.

Mueller, M. L. (2010). *Networks and states: The global politics of Internet governance*. Cambridge, MA: MIT Press.

Musiani, F., Cogburn, D. L., DeNardis, L., & Levinson, N. S. (Eds.). (2016). *The turn to infrastructure in Internet governance*. New York, NY: Palgrave Macmillan.

National Institute of Standards and Technologies (NIST). (2014, February 12). Framework for improving critical infrastructure cybersecurity. Retrieved from https://www.nist.gov/system/files/documents/cyberframework/cybersecurity-framework-021214.pdf

Paré, D. J. (2003). *Internet governance in transition: Who is the master of this domain?* Lanham, MD: Rowman & Littlefield.

Pepitone, J. (2014, January 12). 5 of the biggest-ever credit card hacks. TJX: 94 million. *CNN Business*. Retrieved from https://money.cnn.com/gallery/technology/security/2013/12/19/biggest-credit-card-hacks/3.html

Rousu, J., Saunders, C., Szedmak, S., & Shawe-Taylor, J. (2005, August). Learning hierarchical multi-category text classification models. In *Proceedings of the 22nd international conference on Machine learning* (pp. 744–751). New York, NY: ACM.

Schneider, C. (2016, May 25). The biggest data challenges that you might not even know you have. *IBM Watson*. Retrieved from https://www.ibm.com/blogs/watson/2016/05/biggest-data-challenges-might-not-even-know/

Thierer, A. D., & Crews, C. W. (Eds.). (2003). *Who rules the net? Internet governance and jurisdiction*. Washington, DC: Cato Institute.

Vinton, K. (2014, September 18). With 56 million cards compromised, Home Depot's breach is bigger than Target's. *Forbes*. Retrieved from https://www.forbes.com/sites/katevinton/2014/09/18/with-56-million-cards-compromised-home-depots-breach-is-bigger-than-targets

10   Studying Discourse in Internet Governance through Mailing-List Analysis

Niels ten Oever, Stefania Milan, and Davide Beraldo

Many aspects of contemporary global data flows, including users' ability to enjoy civil liberties online, are shaped by Internet governance processes (DeNardis 2014). Influencing these processes is thus of paramount interest to governments, the industry, and civil society. Engineers and entrepreneurs, lawyers and bureaucrats, and scientists and advocates engage in the development and negotiation of Internet policies and standards in a plethora of fora, each characterized by its own specific configurations of decision-making processes (Hofmann, Katzenbach, and Gollatz 2016). Such a multifaceted scenario results in a wealth of issues, actors, venues, and policy processes that are often intertwined in complex ways (Raboy and Padovani 2010). But it is not just a matter of mere technical details. Because the "arrangements of technical architecture are arrangements of power" (DeNardis 2014, 7), the design of the Internet (Braman 2011) and the associated policy making (Mueller 2002) can be understood as "politics by other means" (Abbate 1999, 179). This makes the study of technical aspects of the Internet and their making, which might otherwise seem solely a matter for engineers, of great interest for social scientists.

Whereas the design, functioning, and decisions of various Internet governance and standard-setting bodies and the participation of different groups have been the topic of several publications (see, among others, DeNardis 2009; Mueller 2010; and Musiani 2013), methodological aspects for the study of Internet governance have received limited attention (e.g., Musiani 2015; Raboy and Padovani 2010). To date, research has relied on discursive methods such as qualitative interviewing and document analysis (e.g., Hintz and Milan 2009; Musiani et al. 2016; Raboy, Landry, and Shtern 2010) or participant observation in policy processes and network analysis (e.g., Hintz 2010; Mueller 2010; Pavan 2012). More recently, however, new software enables automatized analysis, allowing a more granular approach in the study of discursive practices in Internet governance (e.g., Milan and Ten Oever 2017). We argue that software-based tools and methodologies can enhance our understanding of Internet governance and standard-setting processes, in particular with respect to the study of discourse and discursive practices—thus galvanizing this relatively young but swiftly growing field of research.

This chapter explores innovative approaches in the study of discourse within Internet governance settings. Moving from the observation that Internet governance is a "politically contested process of meaning making" (McCharty 2011, 90), we ask what *other* sources of data are available and what can they tell us. What methods are best suited to interrogate these data and processes? While the study of discourse in general remains a crucial focus of Internet governance, we argue that *group discussions* in particular are the natural sites to explore if we are to study the *evolution* of said discourses. In particular, mailing-list archives are a precious and surprisingly underexplored source of data about discursive and norm change as well as stakeholder conflicts and alliances. We contend that only a mixed-methods approach combining computational and interpretative tasks is able to exploit these data sources at their best. In addition, we reflect on the potential of this approach to elicit strategic and tactical interest groups and belonging of social actors, as well as the ethical challenges of this methodological approach. This article tackles some existing challenges to Internet governance scholarship, among those highlighted by DeNardis in chapter 1. In particular, we believe that mailing-list analysis has the potential to contribute to making the invisible visible, by shedding light on otherwise *backstage* decision-making processes and highlighting the inherent power relations. Relatedly, it helps researchers navigating conflicting values, by empowering them to map power coalitions, surface decisional conflicts, and identify marginalized voices.

The chapter is organized as follows. First, we briefly review existing disciplinary and methodological approaches in the study of Internet governance. Second, we define what we mean by "discourse" in the context of the study of Internet governance, largely building on sociological accounts. Third, we explain why mailing lists are a valuable data source, presenting a viable approach to their investigation. We conclude by reflecting on two points: the ethical challenges of the study of mailing lists and the affordances of this approach to support "engaged research" (Milan 2010, 856) decision-making concerning the present and future of crucial infrastructures of our times.

Main Approaches in the Study of Internet Governance

Internet governance increasingly resembles a "mosaic" (Dutton and Peltu 2007), in which relations and issues are knit together. This complex, polycentric ecosystem opens a window on issues of sociopolitical nature that present themselves as technical and of technical issues that turn out to be political (Scholte 2017b). The mosaic itself represents a sort of complex performance (Hofmann 2016) involving a variety of actors—namely, governmental and corporate players, the organized civil society, academia, and the so-called technical community made up of, among others, engineers and computer scientists. A large share of this performance takes place in open bodies, whose functioning and decision-making processes are well documented and publicly accessible. A significant amount of activity, however, still takes place outside public scrutiny (Epstein, Katzenbach, and Musiani 2016).

Developing at the intersection of many processes, Internet governance lends itself to study from a variety of disciplinary perspectives. Each of them brings to the fore specific layers of the mosaic, with implications for the phenomenon that is being observed. This section offers a brief overview of the many layers that can be fleshed out and the distinct disciplinary perspectives that can be adopted when approaching Internet governance as an object of study, bearing in mind that research on Internet governance often focuses on specific areas or issues, with the risk of "equating the overall complexity of the landscape with some of its aspects" (Pavan 2012, xix). Here we review the perspective of those studying the technical and logical layer of the Internet, of those looking at market dynamics subtending the development and operation of the Internet, and of those analyzing the involvement of governances. We conclude by reviewing the holistic approach of science and technology studies (STS).

To start with, Internet governance can be approached from the perspective of computer science in at least three ways. First, computer science plays a vital role in the development of technology supporting the functioning of the Internet, such as communication protocols and other methods of ordering data and data flows. Notable examples are the global database of registered domain holders, WHOIS (Request for Comment [RFC] 3912 [Daigle

2004]), and the registration data access protocol (RFC 7482 [Newton and Hollenbeck 2015]), the latter expected to replace WHOIS and currently being piloted by the Internet Corporation for Assigned Names and Numbers (ICANN).[1] Far from being just a "plumbing matter" (cf. Musiani 2015), these protocols embody distinctive implications of political nature that cannot be understood merely from an engineering perspective. WHOIS, for example, exposes personally identifiable information such as name and home address of the domain holder through a publicly accessible database. At the time of designing the database for what was still a tiny network compared with today's Internet, these implications were not taken into account. Yet the scope of these protocols is far reaching, as testified by the development of Internet protocol version 6 (IPv6) (DeNardis 2009). Second, computer science informs the decisions around the adoption and implementation of technology, because the Internet governance debate heavily relies on computational measures, especially when it comes to the (lack of) adoption of specific standards, protocols, or technologies such as IPv6 (Perset 2010) or the suite of Internet security protocols DNSSEC (Domain Name System Security extensions; Wang 2016). However, while computer science is essential for us to be anchored to the concrete materiality of an issue and its implications, it bears the inherent risk of naturalizing technology, hiding its political implications in the name of the just-because-it-works attitude typical of engineering. Third, and of particular interest here, computer science contributes to the study of Internet governance by developing multipurpose computational methods, ranging from Internet traffic measurements to methods such as those outlined in this chapter (Benthall 2015; Doty 2015; Niedermayer et al. 2016). In this respect, the discipline largely relies on theoretical and experimental methods—whether the development of new hardware, software, communication protocols, algorithms, and databases or the measurement of their effectiveness.

Markets are a key driving force behind the Internet and its governance, thus we ought to consider also economic factors if we are to fully understand its governance arrangements. Since its inception, corporations have played a central role in the development of the Internet—for example, serving as subcontractors for research institutions and the US Department of Defense, which is behind the birth of the Internet as we know it (Frischmann 2001). Their importance radically increased after the privatization of the Internet backbone and the decommissioning in 1995 of the National Science

Foundation Network, a series of US-wide backbone computer networks of the early days (Chinoy and Salo 1997; Kahin 1990). This trend accelerated in the 1990s when the web made the Internet accessible for less-terminal-savvy users. Nowadays, significant levels of market concentration can be observed in all layers of the Internet infrastructure (Dolata and Schrape 2018). Research into market dynamics has thus accompanied the Internet in all these stages, partially because economic policies have played a large role in shaping it (Kahn 1994), and partially because the Internet has a significant impact on the economy (Guillén and Suárez 2005). In Internet governance, considerations of economic nature come into play in the study of scarce resources such as IPv4 addresses (Edelman and Schwarz 2015; Mueller and Kuerbis 2013), the costs of Internet access (Chaudhuri, Flamm, and Horrigan 2005; Prieger 2007), and net neutrality (Greenstein, Peitz, and Valletti 2016; Hahn, Litan, and Singer 2007; Jay and Byung-Cheol 2010). However, these macroeconomic approaches tend to discount the materiality of Internet infrastructure as well its political implications. Moreover, Internet governance is awash with examples of how relevant decisions may happen outside market mechanisms and how control over markets is sometimes pursued via noneconomic means.

Since the early days, when the US government bankrolled the development (Kahn 1994) and supported the global vocation of the Internet, governments have played a significant role in the expansion of the infrastructure. If intergovernmental bodies took a leading role in the coordination of earlier examples of cross-border communication networks such as the telegraph, with the Internet this role has been repeatedly questioned (Chadwick 2006; Drake 2000; Mueller 2010). Communication scholars and political scientists have explored the rise of new governance bodies and how they reconfigure the role of governments (Epstein, Katzenbach, and Musiani 2016), understood through the conceptual lenses of governance innovation (Epstein 2013), regime complex (Nye 2014), and complex hegemony (Scholte 2017a). Works have postulated that governmental regulation might lead to Internet fragmentation (Drake, Cerf, and Kleinwächter 2016; Mueller 2017). While political science, and governance studies and international relations scholars in particular, empowers observers to understand the interactions and the power dynamics involved in Internet governance, as well as the persistent role of governments in them, it embodies a number of limitations. To name just one, governance is a distributed accomplishment that happens not only within governance bodies (Van Eeten and Mueller 2013). Moreover, we cannot understand it as a matter of interaction between only discrete entities—be they states, companies, or civil society actors. As we discuss later, we have to put into focus also the micro interaction layer of individual and small-group participation in decision-making.

Probably the latest addition to the Internet governance tool kit, the discipline of STS has emerged as a particularly fruitful approach. Adopting an STS perspective, scholars have investigated Internet standards as policy documents (Braman 2011), innovation in multistakeholder configurations (Hofmann 2016; Milan and Ten Oever 2016; Ten Oever 2018), the impact of the materiality of infrastructure on the ability of people to exercise their human rights (Cath and Floridi 2017), and infrastructure as a locus of political control (DeNardis and Musiani 2016). STS-inspired approaches allow scholars to weave together both the materially and socially constructed aspects of complex socio-technical processes like Internet governance. Especially through infrastructure ethnography (Bowker et al. 2009; Star 1999) and actor network theory (Latour 2005; Müller 2015), STS can capture the ordering of reality as brought to life by both human and nonhuman actors, as well as the mapping of concrete controversies (Epstein, Katzenbach, and Musiani 2016). However, the focus on the actors' point of view might distract from the bigger picture, obfuscating the role of deeper structures of power and strategic or even deceptive behavior.[2]

While the perspectives and contributions briefly outlined here might appear contiguous yet irremediably apart, we argue that they share a valuable interest for the discursive layer of Internet governance. They variably, and often indirectly, acknowledge that Internet governance is, as McCharty (2011) reminds us, a terrain of political contestation whose object is the construction of meaning associated with infrastructure and society at large. Following McCharty's injunction to take discourse seriously, we now look at what discourse means in Internet governance.

Discourse and Networks in Internet Governance

Discourse gives shape to and reflects the multiple visions and narratives of the Internet as they are developed and advanced by stakeholders. It is in discourse and intra- and intergroup discussion dynamics that meaning making, with its contradictory, chaotic nature instilled in human relations and histories, becomes visible (see Doolin 2003).

Discourse has been under the spotlight of scholars from disciplines as distinct as linguistics, semiotics, and cognitive psychology for as long as since the first half of the twentieth century, thus we will not attempt to provide a comprehensive history of discourse in the sciences. We instead focus on two potential approaches derived from sociology, which posit discourse as at the core of meaning making by and micro interaction between social actors (Melucci 1996) and as something deeply entrenched in the cultures and ideas shaping technological innovation (Flichy 2007).

Combining these perspectives, discourse can thus be seen as the main vehicle for competing values, ideas, and interests to come into focus and play out in multilayered settings by opposed, distinct stakeholders through the contestation over different policy options and technical orderings. It embodies the micro interaction and the narrative dimensions—in which the former includes organizing, mobilization, collaboration, and conflict dynamics, while the latter ranges from beliefs and policy priorities to the "cultural and symbolic understandings surrounding the Internet" (McCharty 2011, 90). Discourse is thus both a vehicle for fostering norm and policy change in Internet policy making and a source of legitimation for the social actors engaged in the process.

As a set of "practices that systematically form the objects of which they speak" (Foucault 1972, 49), discourse can be seen as a locus of power—but contrary to Foucault, the constitution of social relations we are interested in unfolds at the micro level of interaction rather than the macro level of the (historical) social order. As such, discourse is strategically and purposely mobilized by distinct actors (see McCharty 2011). We contend that this perspective can help us capture the multiple levels of contestation that surround Internet policy making and the way ideas and values diachronically evolve, often in surprising ways and unintended directions, through stakeholder interaction. But how can we map the competing narratives that animate, shape, and shake Internet governance arenas? In the next section, we delve into the locus par excellence where discourse unfolds in Internet governance: mailing lists. Studying interaction in mailing lists, we argue, empowers us to investigate qualitatively and quantitatively the discursive formation of policy preferences.

Mailing-List Analysis with BigBang

A distinctive feature of Internet governance bodies such as ICANN and the Internet Engineering Task Force (IETF) is their (relative) openness and the degree of meticulous documentation of their activities through public archives. In most cases, this is not limited to working documents and official outputs but extends to conference calls, public meetings, and cross-community discussions. Everything is recorded and made available on organizational websites for reasons of internal accountability and institutional memory. These archives offer researchers a unique opportunity to investigate the premises of otherwise behind-the-scenes decisions with broad societal implications, thus adding a layer of what we may call external accountability. Besides conference proceedings and documentation, a relevant and convenient repository for this purpose is email archives.

Mailing lists constitute a surprisingly underexplored source of data, holding precious insights on process but also on feelings, values, relationships, and backstage dynamics. The majority of mailing lists are publicly archived, with their archives being publicly accessible to nonmembers of the respective mailing lists. While mailing lists appear to have lost momentum, as today discussion between groups of friends and peers mostly unfolds on messaging apps or social media platforms, they remain a widely used medium in the realm of standards development and other sectors of the Internet governance community. They are extensively used for informal exchange. especially for informal coordination between social actors, and all the way to decision-making, thus making the pathway to decision-making visible. In other words, mailing lists are the locus where the multiple layers that constitute Internet governance sediment and where discourse and discursive practices are enacted in group discussions and collaboration.

Mailing lists offer insights on consensus building and decision-making, conflict and conflict resolution, evolution of a certain issue area and of the language associated with it, group dynamics, power concentration and inclusion or exclusion mechanisms, negotiation tactics, and more. At least four factors make mailing lists a versatile data source with great potential for the study of interactions within the realm of Internet governance:

- Mailing lists are structured. An obvious characteristic of emails is their standardized structure: headers make metadata easily parsable (e.g., by sender, type of interaction, time stamp), which supports different kinds of classification and analysis. Moreover, format standardization facilitates analysis across different mailing lists and the reproducibility of quantitative analysis.

- Mailing lists are cross-sectional. A variety of stakeholder communities engage in discussion—and this all converges and sediments on one or more dedicated mailing lists. Researchers can thus study interaction (e.g., collaboration, contention, and conflict) between interrelated groups and online communities.

- Mailing lists are relational. Data extracted from email archives provide information on the evolving relations among actors (e.g., users' reply chains) or groups (e.g., mailing lists' interlocks), allowing researchers to analyze the structural basis of discourse, power relations among actors, and intergroup dynamics (through, e.g., network analysis).

- Mailing lists are multidimensional. They support social science methods, allowing researchers to circumvent the classical trade-off between scope and depth of the analysis. For instance, simple descriptive statistics (e.g., trends and rankings) can be enriched with relational data (e.g., interaction patterns and user-base overlap), and qualitative textual analysis can be complemented with advanced computational techniques (e.g., machine learning and big data analysis).

- Mailing lists allow longitudinal analysis. They have been the main means for discussing Internet infrastructure and architecture since its inception (see RFC 1155 [Rose and McCoghrie 1990], RFC 1211 [Westine and Postel 1991]). Because the history of mailing lists overlaps with the whole history of the Internet, lists enable a historical approach to Internet-related issues.

BigBang is a Python-based, open source, free software tool kit used by researchers as well as stakeholders in computational and interpretive analysis of mailing lists. At the time of writing, it supports analyzing mailing lists from Sourceforge, Mailman, and .mbox files, among the most common software applications for mailing-list management.[3] Compared with proprietary tools (such as those featured in chapters 2 and 9), open source tools improve verifiability and reproducibility of outcomes and allow more flexibility when adapting the software to specific research challenges.

With BigBang, mailing lists can be analyzed through three main lenses: descriptive statistics, network analysis, and qualitative and quantitative text analysis. Descriptive statistics give us a bird's-eye view about activities in a given mailing list or sets of lists. We can track how many mails have been sent, in what time span, and by how many users. It also allows us to understand the distribution of the length of conversations, commonly known as threads, and the time span in which people were involved in them.

Network analysis of mailing lists reveals patterns of communications, their development, and the role senders have in the community. It helps in understanding whether certain participants function as a node for the dissemination of ideas across mailing lists and communities or whether conversations stop when certain actors (or groups of actors) get involved. It also shows the centrality of actors and their proximity to other nodes and whether early exchanges could, for instance, be indicators of emerging relations by reflecting an increase in shared messages over time. Finally, it reveals the distribution of individuals and groups (and subgroups) within the larger landscape and the connections between them: who talks to whom and who are the connectors across distinct stakeholder groups.

With qualitative and quantitative text analysis, we can combine the descriptive statistics and network analysis and ask questions with the two, moving past a basic question like what are the trending topics in conversations. It also allows us to analyze the affiliation of participants (Niedermayer et al. 2016), which in turn aids investigation of the role of formal and informal leaders in online communities. The combination of affiliation and formal and informal leadership roles helps in analyzing the responses of structured groups to specific topics and patterns. Because mailing lists allow text analysis one can couple the study of mailing lists with the analysis of other structured text, such as contributions to the code repository GitHub, policy documents, membership or participation registries, statements of interest, and meeting transcripts. This then can be used to investigate how affiliation, gender, RFC authorship, or other characteristics relate to levels of participation, the mode of participation, patterns in responses, etcetera. Finally, the computational analysis of mailing lists can offer pointers for more interpretative approaches such as Foucauldian, or critical, discourse analysis.

Like other approaches, quantitative mailing-list analysis in general, and BigBang more specifically, comes with some caveats. First, data sources typically contain biases, are incomplete, or are even systematically flawed (Karpf 2012), and mailing lists are no exception. Regarding data accuracy, we cannot but note that some dynamics, such as the presence of passive members, are not made visible by interaction in mailing lists, which archive only mails that were sent to the list. Furthermore, only those mailing lists whose existence is known to the researcher can be analyzed—and only if the researcher has access to the archives. Some mailing lists do not hold archives; others are not public at all or do not allow public subscription. Not all data are correctly captured: in our analysis of bulk data from ICANN and IETF mailing lists, for example, we have come across emails erroneously dated as far back as 1904, and with otherwise obviously wrong timestamps (e.g., "32 Jan 2008")—and these were the ones we were able to identify and filter out. Occasionally, mailing-list archives contain spam, which might alter the results—but BigBang offers options for filtering spam out of the archives.

Conclusion

In this chapter we explore the added value of mailing-list analysis as a venue for investigating micro interaction and narratives in Internet governance and illustrate the potential of the BigBang tool kit. However, while the computerized analysis of interaction in mailing lists represents a fruitful venue to study discourse and discursive practices at the micro interaction level, there are some ethical considerations researchers need to attend to in particular with respect to privacy, anonymity, and consent. Although the mailing lists we used for our research are publicly archived, analyzing discourse and discursive practices may, for example, offer additional keys to understanding aspects of in-group interaction that might jeopardize group activities or dangerously single out certain users. Anonymizing utterances in publicly archived lists is impossible, as a search of public archives by third parties would easily reveal the author and other important metadata. Consent, then, is hard to obtain from every single participant but might be easier to obtain from an organization. Consent might come through the terms and conditions that come with mailing-list subscriptions or the expectations one might have when participating in a governance forum through its official channels, as happened in our research on ICANN and IETF lists. At least two questions arise: Is this a sufficient safeguard? What about "group privacy" (Taylor, Floridi, and Van der Sloot 2017)? In sum, with mailing-list analysis there is no one-size-fits-all ethical approach, and researchers can expect to have to make ad hoc considerations with respect to the participants' reasonable expectations in public mailing lists in a given sociopolitical context.

We conclude by exploring the claim that the study of (Internet) governance influences governance dynamics and outcomes. As Ziewitz and Pentzold put it,

> Given the role of governance research in rationalising, justifying and legitimating political interventions, methods cannot be viewed as neutral instruments. Interestingly, however, questions of methodology are only rarely discussed in studies of Internet governance. Most studies still tend to rely on case studies that are largely presented as unproblematic representations of reality, which are not further questioned in the course of the analysis. The absence of such methodological reflection makes sense in that it contributes to the performativity of governance by not inducing the reader to question the text and its authority. (2014, 318)

We believe mailing-list analysis with BigBang can be repurposed as a tool for engaged research; that is to say, an approach to inquiry that, without departing from systematic, evidence-based social science research, may support the attempts of advocates to set the agenda of policy makers (Milan 2010, 856). It can, for example, improve the accountability of actors and stakeholder groups engaged in Internet governance and uncover allies and alignments on specific policy options and technical orderings, perhaps advancing social concerns in the Internet governance landscape.

Acknowledgments

This project has received funding from the European Research Council (ERC) under the European Union's Horizon 2020 research and innovation program (grant agreement No. 639379-DATACTIVE, awarded to Stefania Milan as principal investigator; https://data-activism.net).

Notes

1. Requests for Comment (RFC) are the output documents of the Internet Engineering Task Force, the Internet Architecture Board, and the Internet Research Task Force pertaining to Internet infrastructure topics (see also Ten Oever and Moriarty 2018).

2. For a more comprehensive discussion of the role of STS, see chapter 3, in which Musiani thoroughly analyzes the contribution of STS to the study of Internet governance.

3. See the BigBang website at https://github.com/datactive/bigbang.

References

Abbate, J. (1999). *Inventing the Internet: Inside technology*. Cambridge, MA: MIT Press.

Benthall, S. (2015). Testing generative models of online collaboration with BigBang. *Proceedings of the 14th Python in Science Conference* (pp. 182–189). Retrieved from https://conference.scipy.org/proceedings/scipy2015/sebastian_benthall.html

Bowker, G. C., Baker, K., Millerand, F., & Ribes, D. (2009). Toward information infrastructure studies: Ways of knowing in a networked environment. In J. Hunsinger, L. Klastrup, & M. M. Allen (Eds.), *International Handbook of Internet Research* (pp. 97–117). Dordrecht, Netherlands: Springer.

Braman, S. 2011. The framing years: Policy fundamentals in the Internet design process, 1969–1979. *The Information Society, 27*(5), 295–310.

Cath, C., & Floridi, L. (2017). The design of the Internet's architecture by the Internet Engineering Task Force (IETF) and human rights. *Science and Engineering Ethics, 23*(2), 449–468.

Chadwick, A. (2006). *Internet politics: States, citizens, and new communication technologies*. New York, NY: Oxford University Press.

Chaudhuri, A., Flamm, K. S., & Horrigan, J. (2005). An analysis of the determinants of Internet access. *Telecommunications Policy, 29*(9), 731–755.

Chinoy, B., & Salo, T. J. (1997, September). Internet exchanges: Policy-driven evolution. In B. Kahin & J. H. Keller (Eds.), *Coordinating the Internet* (pp. 325–345). Cambridge, MA: MIT Press.

Daigle, L. (2004). *WHOIS protocol specification*. RFC 3912. Retrieved from Internet Engineering Task Force website: https://tools.ietf.org/html/rfc3912

DeNardis, L. (2009). *Protocol politics: The globalization of Internet governance*. Cambridge, MA: MIT Press.

DeNardis, L. (2014). *The global war for Internet governance*. New Haven, CT: Yale University Press.

DeNardis, L., & Musiani. F. (2016). Governance by infrastructure: Introduction. In F. Musiani, D. L. Cogburn, L. DeNardis, & N. S. Levinson (Eds.), *The turn to infrastructure in Internet governance* (pp. 3–21). Basingstoke, UK: Palgrave Macmillan.

Dolata, U., & Schrape, J. F. (2018). *Collectivity and power on the Internet: A sociological perspective*. London, UK: Springer.

Doolin, B. (2003). Narratives of change: Discourse, technology and organization. *Organization, 10*(4), 751–770.

Doty, N. (2015). Reviewing for privacy in Internet and web standard-setting. In *Security and Privacy Workshops (SPW), 2015 IEEE* (pp. 185–192). San Jose, CA: IEEE.

Drake, W. J. (2000). Rise and decline of the international telecommunications regime. In C. Marsden (Ed.), *Regulating the global information society* (pp. 124–177). London, UK: Routledge.

Drake, W. J., Cerf, V. G., & Kleinwächter, W. (2016, January). Internet fragmentation: An overview. Retrieved from World Economic Forum website: http://www3.weforum.org/docs/WEF_FII_Internet_Fragmentation_An_Overview_2016.pdf

Dutton, W. H., & Peltu, M. (2007). The emerging Internet governance mosaic: Connecting the pieces. *SSRN*. Retrieved from https://doi.org/10.3233/IP-2007-0113

Edelman, B., & Schwarz, M. (2015). Pricing and efficiency in the market for IP addresses. *American Economic Journal: Microeconomics, 7*(3), 1–23.

Epstein, D. (2013). The making of institutions of information governance: The case of the Internet Governance Forum. *Journal of Information Technology, 28*(2), 137–149.

Epstein, D., Katzenbach, C., & Musiani, F. (2016). Doing Internet governance: Practices, controversies, infrastructures, and institutions. *Internet Policy Review, 5*(3). Retrieved from https://doi.org/10.14763/2016.3.435

Flichy, P. (2007). *The Internet imaginaire*. Cambridge, MA: MIT Press.

Foucault, M. (1972). *The archaeology of knowledge*. New York, NY: Harper & Row.

Frischmann, B. (2001). Privatization and commercialization of the Internet infrastructure. *Columbia Science & Technology Law Review, 2*, 1–25.

Greenstein, S., Peitz, M., & Valletti, T. (2016). Net neutrality: A fast lane to understanding the trade-offs. *Journal of Economic Perspectives, 30*(2), 127–150.

Guillén, M. F., & Suárez, S. L. (2005). Explaining the global digital divide: Economic, political and sociological drivers of cross-national Internet use. *Social Forces, 84*(2), 681–708.

Hahn, R. W., Litan, R. E., & Singer, H. J. (2007). The economics of "wireless net neutrality." *Journal of Competition Law & Economics, 3*(3), 399–451.

Hintz, A. (2010). *Civil society media and global governance: Intervening into the World Summit on the Information Society*. Berlin, Germany: LIT Verlag.

Hintz, A., & Milan, S. (2009). At the margins of Internet governance: Grassroots tech groups and communication policy. *International Journal of Media & Cultural Politics, 5*(1/2), 23–38.

Hofmann, J. (2016). Multi-stakeholderism in Internet governance: Putting a fiction into practice. *Journal of Cyber Policy, 1*(1), 29–49.

Hofmann, J., Katzenbach, C., & Gollatz, K. (2016). Between coordination and regulation: Finding the governance in Internet governance. *New Media & Society, 19*(2), 1406–1423.

Jay, P. C., & Byung-Cheol, K. (2010). Net neutrality and investment incentives. *The RAND Journal of Economics, 41*(3), 446–471.

Kahin, B. (1990). *Commercialization of the Internet: Summary Report*. RFC 1192. Retrieved from Internet Engineering Task Force website: https://tools.ietf.org/html/rfc1192

Kahn, R. E. (1994). The role of government in the evolution of the Internet. *Communications of the ACM, 37*(8), 15–19.

Karpf, D. (2012). Social science research methods in Internet time. *Information, Communication & Society, 15*(5), 639–661.

Latour, B. (2005). *Reassembling the social: An introduction to actor-network-theory*. Oxford, UK: Oxford University Press.

McCharty, D. R. (2011). Open networks and the open door: American foreign policy and the narration of the Internet. *Foreign Policy Analysis, 7*, 89–111.

Melucci, A. (1996). *Challenging codes. Collective action in the information age*. Cambridge, UK: Cambridge University Press.

Milan, S., & ten Oever, N. (2017). Coding and encoding rights in Internet infrastructure: Sociotechnical imaginaries and grassroots ordering in Internet governance. *Internet Policy Review, 6*(1). Retrieved from https://doi.org/10.14763/2017.1.442

Mueller, M. (2002). *Ruling the root: Internet governance and the taming of cyberspace*. Cambridge, MA: MIT Press.

Mueller, M. (2010). *Networks and states: The global politics of Internet governance*. Cambridge, MA: MIT Press.

Mueller, M. (2017). *Will the Internet fragment? Sovereignty, globalization and cyberspace*. New York, NY: John Wiley & Sons.

Mueller, M. L., & Kuerbis, B. (2013). Buying numbers: An empirical analysis of the IPv4 number market. In *iConference 2013 Proceedings* (pp. 27–37). Retrieved from https://www.ideals.illinois.edu/handle/2142/36045

Müller, M. (2015). Assemblages and actor-networks: Rethinking socio-material power, politics and space. *Geography Compass, 9*(1), 27–41.

Musiani, F. (2013). *Nains sans géants. Architecture décentralisée et services Internet*. Paris, France: Presses de Mines.

Musiani, F. (2015). Practice, plurality, performativity, and plumbing: Internet governance research meets science and technology studies. *Science, Technology & Human Values, 40*(2), 272–288.

Musiani, F., Cogburn, D. L., DeNardis, L., & Levinson, N. S. (2016). *The turn to infrastructure in Internet governance*. Basingstoke, UK: Palgrave Macmillan.

Newton, A., & Hollenbeck, S. (2015). *Registration Data Access Protocol (RDAP) query format*. RFC 7482. Retrieved from Internet Engineering Task Force website: https://tools.ietf.org/html/rfc7482

Niedermayer, H., Raumer, D., Schwellnus, N., Cordeiro, E., & Carle, G. (2016). An analysis of IETF activities using mailing-lists and social media. In *Internet Science*, 218–230. Lecture Notes in Computer Science. Cham, Switzerland: Springer.

Nye, J. S. (2014). *The regime complex for managing global cyber activities*. Global Commission on Internet Governance Paper Series (Paper no. 1). Centre for International Governance Innovation/Chatham House. Retrieved from https://www.cigionline.org/sites/default/files/gcig_paper_no1.pdf

Pavan, E. (2012). *Frames and connections in the governance of global communications: A network study of the Internet Governance Forum*. Lanham, MD: Lexington Books.

Perset, K. (2010, April). Internet addressing: Measuring deployment of IPV6. Retrieved from OECD website: https://doi.org/10.1787/5kmh79zp2t8w-en

Prieger, J. E. (2007). The supply side of the digital divide: Is there equal availability in the broadband Internet access market? *Economic Inquiry, 41*(2), 346–363.

Raboy, M., Landry, N., & Shtern, J. (2010). *Digital solidarities, communication policy and multi-stakeholder global governance*. New York, NY: Peter Lang.

Raboy, M., & Padovani, C. (2010). Mapping global media policy: Concepts, frameworks, methods. *Communication, Culture & Critique, 3*(2), 150–169.

Rose, M., & McCoghrie, K. (1990). *Structure and identification of management information for TCP/IP-based Internets*. RFC 1155. Retrieved from Internet Engineering Task Force website: https://tools.ietf.org/html/rfc1155

Scholte, J. A. (2017a, December). *Complex hegemony: The IANA transition in global Internet governance*. Paper presented at the 12th Annual Symposium of the Global Internet Governance Academic Network (GigaNet), Geneva, Switzerland.

Scholte, J. A. (2017b). Polycentrism and democracy in Internet governance. In U. Kohl (Ed.), *The net and the nation state: Multidisciplinary perspectives on Internet governance* (pp. 165–184). Cambridge, UK: Cambridge University Press.

Star, S. L. (1999). The ethnography of infrastructure. *American Behavioral Scientist, 3*, 377–391.

Taylor, L., Floridi, L., & van der Sloot, B. (Eds.). (2017). *Group privacy: New challenges of data technologies*. London, UK: Springer.

ten Oever, N. (2018). Productive contestation, civil society, and global governance: Human rights as a boundary object in ICANN. *Policy & Internet, 11*(1), 37–60.

ten Oever, N., & Moriarty, K. (Eds.). (2018). The tao of IETF: A novice's guide to the Internet Engineering Task Force. Retrieved from Internet Engineering Task Force website: https://www6.ietf.org/tao

van Eeten, M. J., & Mueller, M. L. (2013). Where is the governance in Internet governance? *New Media & Society, 15*(5), 720–736.

Wang, Z. (2016, February). Take up DNSSEC when needed. Retrieved from http://arxiv.org/abs/1602.08459/

Westine, A., & Postel, J. (1991). *Problems with the maintenance of large mailing lists.* RFC 1211. Retrieved from Internet Engineering Task Force website: https://tools.ietf.org/html/rfc1211

Ziewitz, M., & Pentzold, C. (2014). In search of Internet governance: Performing order in digitally networked environments. *New Media & Society, 16*(2), 306–322.

11   The Biases of Information Security Research

Ronald J. Deibert

Information and communication technologies now pervade, and are criti-cal to the functioning of, every activity on earth.[1] Securing information technologies has, as a consequence, become a major public policy issue. But *how* to secure information and communication technologies is deeply contested (Deibert 2017). The choices that are made to secure information ultimately shape our communications spaces and, by extension, how we secure our societies.

Research on information security is, therefore, critical to navigating this issue in an informed and systematic way. But what is a problem in informa-tion security that should be addressed by research in the first place? On first consideration it may seem obvious that the security of information systems is a technological issue and that eventually, with enough systematic atten-tion to engineering of information systems, we can settle on a way to secure information that achieves consensus.

But approaching information systems through a technical-functional lens obscures the politically contested nature of information security as well as how research on information security itself is influenced by external factors. In this chapter, I outline some of the ways in which information security gets defined as a research problem. My aim is to show how politi-cal, economic, and other factors shape the field of information security, including what problems are excluded as much as which are addressed or prioritized. After this analysis, it will be clearer how the field of information security is constructed as a regime or community of practice and the impli-cations this shaping has for society and politics writ large (Adler 2008).

In emphasizing these questions, I am not trying to solve technological problems that plague information security. Rather, I am asking epistemo-logical and methodological questions about the constitution of the field of information security as it exists today. As Langdon Winner puts it, "It is important to notice not only which decisions are made and how but also which decisions never land on the agenda at all; which possibilities are relegated to the sphere of non-decisions" (Winner 1993, 369). Answering these questions requires studying the past, understanding how the field of information security evolved and where it is headed today, including a consideration of how political and economic forces shape what counts as legitimate areas of inquiry and, as importantly, what does not. In short, it requires a sociology of information security knowledge (see Bourdieu 2004; Pinch and Bijker 1993).

I conclude with some suggestions on how to make research on information security more self-aware, more independent, and more broadly interdisciplinary—values that I hope help militate against biases and unquestioned assumptions.

The Barriers of Disciplinary Silos

A useful starting point for this inquiry is a reminder that academic fields of inquiry are, themselves, social constructions. These fields are made up of communities of practice that share common principles, rules, languages, methods, and even worldviews. These common traits do not mean that all academics within a particular field or discipline think alike. To the contrary, academic fields—especially those that are healthy—are typically characterized by rigorous debates between competing schools of thought. Notwithstanding their internal differences, however, disciplines create fences around research, commonly referred to as disciplinary silos, which have the effect of channeling inquiry into conventions. These conventions are necessary adaptations to a complex world, which requires specialization and division of labor. The downside of disciplinary silos is that they can encourage myopic thinking while discouraging alternative ways of approaching a topic.

With respect to the field of information security research, the most important disciplinary silo is the one separating the computer and engineering sciences from the social sciences and humanities. Few universities today systematically include engineering and computer science curricula in the social sciences, and vice versa. These disciplinary divisions are compounded by the growing imbalance between the natural sciences, on the one hand, and the social sciences and humanities on the other. Research into science, technology, engineering, and mathematics (STEM) subjects is growing, mostly on the basis of economic arguments. STEM subjects are considered pathways to employment, innovation, and economic growth, while the social sciences and many humanities are increasingly portrayed as frivolous and financially unrewarding. Whereas STEM subjects receive the majority of government and private sector funding, humanities programs are shrinking or being eliminated altogether in absence of equivalent funding. These trends reinforce existing disciplinary divisions in information security research while also privileging technical approaches to information security problems over broader political and economic considerations.

To be sure, there are exceptions. There are growing subfields dealing with human-computer interaction and user interface design, which draw insights and theories from psychology and sociology to better design information systems. Many social scientists and humanists use methods that rely on computer science techniques for quantitative and qualitative data analysis (see chapter 9 by Cogburn and chapter 10 by Ten Oever, Milan, and Beraldo). Information studies programs are growing into a self-identified iSchool movement in which the academic field of information is envisioned as an interdisciplinary whole. Many iSchools include faculty undertaking research on topics broadly related to information security, such as censorship and surveillance.

Another important exception is the subfield of security economics, spearheaded by Cambridge University professor Ross Anderson (n.d.) and others. Anderson and his growing cohort of colleagues examine the incentive structures that shape information security practices, particularly within software and other technology industries. Since most of the IT sector is self-regulated, with competitive dynamics rewarding early innovators, companies have few incentives to invest in protection from the ground up, meaning that those responsible for producing information systems pass on the costs of security to the end user or other parties. As Anderson (2014) explains, "Information economics explain why security is hard: in platform races, winners are likely to be those who ignored security to make life easy for complementers, such as application developers, and they are then likely to use it to lock customers in later rather than just protecting them from the bad guys."

While Anderson's and his colleagues' work bridges STEM and the social sciences, it is incomplete. Although it incorporates economic theory into an analysis of information security outcomes, it does not examine how political and economic factors may shape what counts as legitimate research into information security in the first place. In other words, it does not extend the analysis of positive and negative incentives into critically examining and reflecting on why the field privileges certain areas of inquiry and methods over others, and what that privileging means for how, and in whose interests, information is secured.

These examples of interdisciplinarity, while encouraging, are still exceptions to strong and well-established conventions. Moreover, there are growing structural constraints that could bias research in ways that reinforce not only those disciplinary conventions but also dominant interests. We turn now to an analysis of some of those structural constraints.

Structural Constraints on Information Security

Langdon Winner once wrote that "by noticing which issues are never (or seldom) articulated or legitimized, observing which groups are consistently excluded from power, one begins to understand the enduring social structures upon which the most obvious kinds of political power rest" (Winner 1993, 369). What do we notice when we apply this framework to information security? My and my colleagues' research on targeted digital threats at the Citizen Lab provides a compelling example (Citizen Lab 2014). At the Citizen Lab, we have analyzed digital espionage campaigns against civil society groups since 2009. Beginning with our contribution to the landmark *Tracking GhostNet* report (Information Warfare Monitor 2009), and ever since, we have documented persistent cyber espionage campaigns targeting civil society, human rights defenders, political opposition, journalists, and even health scientists. Our conclusion is that the same threat actors—typically government agencies or nonstate actors working on behalf of states—use the same tactics, techniques, and procedures against civil society that they use in espionage campaigns against private sector and government targets. However, civil society typically lacks the resources, support, and capacity to deal with the threat. Defending against digital espionage, from investigation to ongoing remediation, can involve huge investments in money, time, equipment, and personnel that few civil society groups can afford.

Compounding the problem, civil society is not a lucrative client for the multibillion-dollar cybersecurity threat industry. As a consequence, even though private firms may identify civil society targeting in their monitoring data, there are few fiscal incentives for these companies to follow up and even some potential risks in doing so. While their reporting may list generalized nongovernmental organization (NGO) or civil society targets and victims, those are rarely analyzed in depth. Instead, the reports principally aim to highlight industry and government targets and victims, which are more attractive sources of revenue. These industry reports can then perpetuate misguided assumptions about the actual scope and scale of information security problems, about who is targeted in digital espionage, and why (Maschmeyer, Deibert, and Lindsay 2019; see also Jardine's chapter 7 for other biases in cybersecurity datasets introduced by the private sector).

Meanwhile, a variety of factors militate against policy makers addressing digital espionage against civil society. For many nondemocratic governments, and especially authoritarian regimes, civil society can represent a menace to be co-opted, infiltrated, and neutralized. These governments may see civil society as an information security threat. But even in democratic countries there are few incentives to address this problem. For example, in a 2014 case, a US grand jury indicted five Chinese individuals for "computer hacking, economic espionage and other offenses directed at six American victims in the US nuclear power, metals and solar products industries" (US Department of Justice 2014). Many link the indictment to a subsequent agreement between US President Barack Obama and China's President Xi Jinping to refrain from computer espionage targeting businesses and critical infrastructure (Harold 2016). It is unlikely that equivalent political capital would be marshalled for such an indictment in support of civil society targets. The theft of trade secrets affects jobs, and national security is an existential threat that attracts bipartisan support, but the targeting of Tibetans, human rights NGOs, or journalists does not garner the same political urgency.

This example illustrates how the structures Winner refers to can obscure an important area of information security, in this case the silent epidemic of targeted digital attacks against civil society. Targeted digital attacks on civil society groups tend to be underreported and largely marginalized because, for the private sector cybersecurity threat industry, defending those attacks is not profitable and investments of time or resources cannot easily be justified. For government agencies, the consequences of such attacks rarely rise to the level of a political or economic risk that merits public policy attention. Indeed, a growing number of governments actively target the sector.

Meanwhile, civil society has very little capacity to deal with the problem, creating what is, overall, a failure in the information security market.

A large body of academic research on this area, combined with persistent political advocacy directed at governments and the threat industry, could raise awareness and help address that market failure, which is precisely one of the aims of our research. However, the hurdles standing in the way are numerous and growing.

A Political Economy of Information Security Research

In a world of billions of networked devices, it may be easy to forget that the Internet began as a relatively small university-based research project. Contrary to the widespread myth of the Internet's origins linking Paul Baran's proposal for a distributed communications system that could survive a nuclear war, the early networking infrastructures that were developed and that evolved into today's Internet were concerned with facilitating time-sharing of scarce computing resources, not just surviving nuclear attack. Engineers focused on enabling network connectivity across nodes and having data flow seamlessly. Security of that network or its communications was not a high priority (Abbate 1999).

It was not until the Morris worm, in 1988, that the threats around network failure and insecurity became more widely noticed and increasingly studied. Researchers began examining how poorly executed code could cause systems to crash or be exploited to cause havoc. Significantly, the lens through which the growing subfield of information security addressed their problems was from a technical-functional one. Research was aimed at highlighting technical weaknesses in information security systems' design to optimize against failure, whether accidental or otherwise.

A technical-functional approach may have been appropriate when information systems were relatively obscure and compartmentalized. But the contemporary information ecosystem, which extends into cultural and political systems across the world, is vastly different from that in which the early innovators worked. The different institutional environments affect what counts as an information security problem in the first place, making the security of information less a *technical* than a *political* issue. Decisions on what problems to resolve, and which issues are not even regarded as problems, are questions of *power*; once those questions are resolved, technical and engineering questions follow along. Practices of information security, seen as a component of technological production, are thus exercises in the architecture of power. Starting from this perspective can help illuminate what gets priority as an information security research problem and what does not.

Economic Biases of Information Security Research

The problems for which engineering and computer science solutions are sought and priorities allocated may reflect factors such as the progress of debates in a particular academic field or the evolution of technologies themselves. But they may also reflect the existing power structure of a society and the constraints and opportunities it presents, both directly and indirectly. Among power structures today, economic factors have enormous influence on a wide variety of sectors, academic research among them. For example, it has become increasingly commonplace for universities to justify their programs in terms of an investment in or contribution to the economy. Research programs sell themselves in terms of employment training. This dynamic has positive and negative incentives on research. Positive incentives push for certain problems to be addressed because they are problems for which dominant industries seek creative solutions. For example, social media companies' appetites for better ways of pushing advertisements to their customers on the basis of their online behavior undoubtedly incentivizes researchers to find more efficient ways to monitor users. These incentives can be direct, through financial support, or indirect, in terms of potential postgraduate employment opportunities.

Negative incentives dissuade researchers from looking into certain areas of inquiry because doing so may jeopardize those employment opportunities or risk some kind of legal or other penalty. Examples include undertaking research into security flaws or privacy violations associated with large prospective employers. Who wants to interrogate a hidden and prejudiced algorithm if it will invoke the wrath of a company's powerful legal department or blacklist graduates from prospective jobs?

Negative incentives may affect the choice of research methods as well. For example, reverse engineering software—a method used widely to undertake research on software—may infringe on copyright or intellectual property laws, such as section 1201 of the US Digital Millennium Copyright Act or the US Computer Fraud and Abuse Act. Such laws effectively shackle certain modes of inquiry by shielding code and networks from outside scrutiny. While protecting business secrets may be the official justification for keeping code hidden, there may be other, political, reasons as well.

For example, in numerous investigations our research group has shown that many popular China-based applications contain hidden censorship and surveillance mechanisms that the developers have engineered to comply with their understanding of government regulations. The research provides an invaluable insight into China-based information controls on popular chat, video-streaming, and gaming applications and how user security and privacy is subordinated to comply with regime security. But none of the research would be possible without our researchers reverse engineering the applications, in some cases breaking their encryption, to discover what algorithms lie beneath the surface. A strict interpretation of intellectual property laws could chill this research, especially if the companies under scrutiny are litigious.

These concerns are not just hypothetical. Our group has been the subject of legal threats in at least three instances by companies who were the subjects of our research. In 2014, the security company Hacking Team experienced a massive data breach. Those responsible leaked the stolen data to WikiLeaks, where it was published online in searchable form. Citizen Lab had published several reports on how the company's products were used by autocratic regimes to target human rights defenders, journalists, and members of political opposition parties; these reports were widely covered in the media and generated negative publicity for the company. A search of their corporate emails as archived on WikiLeaks showed that executives reviewed the letters we sent to the company before we published reports, in which we asked questions about company practices, and that they engaged formal legal advice to "hit CL [Citizen Lab] hard" (Lou 2015).

As a component of research into a deep packet inspection system manufactured by the company Sandvine and its use in Turkey, Syria, and Egypt to redirect unwitting users to spyware and other malicious traffic, Citizen Lab researchers purchased a secondhand Sandvine PacketLogic device from an online commerce site (Marczak et al, 2018). We set up the device in a laboratory to understand its operation and refine the fingerprints we used as part of our network measurement methodology. Before publication, we sent Sandvine a letter describing our research and asking questions about its use in the country contexts. Sandvine threatened legal action against the University of Toronto if the device was not immediately returned. The University of Toronto denied the request, and the publication proceeded in spite of the threat of legal action.

While Hacking Team and Sandvine did not actually launch legal action, another company did. Netsweeper became notorious after several Citizen Lab reports showed that the company's Internet filtering technology was enabling national-level censorship in several countries, including of political and social content. The company filed a $3.5 million defamation suit against me, as director of the Citizen Lab, and the University of Toronto, over comments made to the media about a Citizen Lab report on Netsweeper's services to Yemen. Canada has historically proved a plaintiff-friendly environment for defamation cases. However, on November 3, 2015, the legal landscape shifted in Ontario when a new law called the Protection of Public Participation Act (PPPA) came into force. It was specifically designed to limit strategic litigation against public participation, or SLAPP, suits. In our view, the work of Citizen Lab, including the research on Netsweeper, was precisely the sort of activity recognized as meriting special protection under the PPPA. Had our proceedings gone forward, we intended to exercise our rights under the act and move to dismiss Netsweeper's action. Ultimately, however, Netsweeper discontinued its defamation suit in its entirety. Regardless, the case underscores the threats to controversial security research that can come from the private sector as well as the types of legal remedies that may protect researchers who engage in that controversial research. I return to this topic in the conclusion.

Private Sector Gatekeepers on Information Security Research

Our information environment is mediated by large technology companies. These companies control massive repositories of data that are relevant to information security issues, from protection of privacy to censorship and surveillance, to the circulation of state disinformation campaigns, or to efforts to further violent radical extremism. They are also extraordinary potential opportunities as datasets for researchers to interrogate. Indeed, going further than describing these as mere opportunities, it is perhaps better to describe access to such data as a research *imperative*. The less researchers know about how these data are collected, analyzed, produced, and used to shape and limit users' communications experiences, the less they can authoritatively claim to know about what are arguably some of the most important information security issues of the day.

But what are the conditions of access? Who can access the data and under what terms? As Lev Manovich (2011, 464) notes, "Only social media companies have access to really large social data—especially transactional data. An anthropologist working for Facebook or a sociologist working for Google will have access to data that the rest of the scholarly community will not." Apart from companies publishing only data or research they believe is in line with their for-profit aims, this privatization of core sociological research data may have several consequences for information security research. As boyd and Crawford explain,

> Some companies restrict access to their data entirely; others sell the privilege of access for a high fee; and others offer small data sets to university-based researchers. This produces considerable unevenness in the system: those with money—or those inside the company—can produce a different type of research than those outside. Those without access can neither reproduce nor evaluate the methodological claims of those who have privileged access. (2011, 22)

Researchers who want to work on datasets controlled by companies have some options. Some companies provide access to portions of the data, and many researchers have made use of this access to undertake pathbreaking research. But typically this access is restricted to slices of data or comes with subscription fees to larger, or even complete, datasets. Segregating access to data in this way reinforces all the usual resource inequities that are found across societies: better-resourced departments and universities acquire more and better access than those that are less endowed. This division reinforces long-standing inequities not only across disciplines but across the world, and in particular for researchers in the global South.

The desire to access data may also influence the topics on which researchers choose to focus. Companies may encourage access on the basis of having researchers tackle problems that companies need to solve and, conversely, discourage questions that affect them adversely. As boyd and Crawford (2011, 674) point out, "Big Data researchers with access to proprietary data sets are less likely to choose questions that are contentious to a social media company…if they think it may result in their access being cut."

The influence of private sector and economic factors biasing information security research is likely to grow, as industry becomes a more important sponsor of university-based research. According to a report from the US

National Science Foundation, from 2011 to 2014 federal funds for university research fell to $37.9 from $40.8 billion, while during the same period industry-sponsored university research grew from $4.9 to $5.9 billion.[2]

Although biases introduced into university-based research from the private sector are definitely not unique to information security, the area does have special considerations that bear on the topic. All things being equal, large social media companies and other large technology providers would benefit by a more narrowly defined version of security that focuses largely on technical matters. Companies whose business model rests on surveillance of users' online behavior are unlikely to sponsor research that undermines that model or helps users become aware of just how much they are giving away. These incentives privilege a certain narrow definition of what counts as information security, one in which user privacy is not included.

Companies will also not likely look favorably on research that highlights collusion with governments on lawful access requests or implementation of surveillance or censorship of users, which unfortunately is a growing norm among countries worldwide (Deibert and Crete-Nishihata 2012). A comparison of the full range of state cybersecurity policies would show fairly wide variation in what counts as requiring securing and what practices are facilitated or encouraged by the policies that are adopted. National-level policies have knock-on effects: on companies, consumer behavior, and to a large degree, academic research. For example, China's cybersecurity laws require companies to police their networks, censor forbidden topics, monitor their users, and store their customer data on China's territory, to be available to authorities upon request. Naturally, the laws affect how products are designed, engineered, and effectively secured. They incentivize engineering puzzles to be solved and priorities to be addressed. Research resources are directed accordingly: perhaps to find ways to better monitor users' devices and online behavior or develop ways to restrict their access to information in a manner that is increasingly opaque to users and researchers.

An intriguing and disturbing example concerns research into artificial intelligence (AI). The Chinese government's "Next Generation Artificial Intelligence Development Plan" links AI to China's so-called social credit system—a system whose aim is to combine a comprehensive analysis of users' online behavior with a kind of citizenship score. Major Chinese social media companies, like Baidu and Tencent, all of which engineer extensive censorship and surveillance into their products, have invested heavily in

AI research labs. A 2017 article in *The Atlantic* describes one Hong Kong University–based researcher's collaboration with the company Tencent:

> The students get access to mountains of data from WeChat, the messaging app from Tencent that is akin to Facebook, iMessage, and Venmo all rolled into one. ("With AI, they can't do it without a lot of data and a platform to test it on," says Yang, which is why industry collaboration is so key.) In return, Tencent gets a direct line to some of the most innovative research coming out of academic labs. And of course, some of these students end up working at Tencent when they graduate. (Zhang 2017)

As government pressures to control information bear down heavily on the private sector, both positive and negative incentives faced by firms are invariably passed down to research communities, particularly as industry support for university research increases. On balance, these pressures work to squeeze out spaces for information security research on censorship, surveillance, lawful access, and privacy.

National Security Biases

The control of information has long been a central ingredient of state power. States have used information to control their domestic populations or project power abroad, through forms of propaganda or information operations. Today, the global information environment sees struggles between states, as well as nonstate actors such as private companies, NGOs, and violent extremists. Within this multifaceted arena of struggle, numerous practices of information security, in turn, can put pressure on and shape information security research. While the previous section highlights some indirect influences of national security via the private sector, in this section we examine some direct impacts.

The most important direct tension between national security and university-based information security research is that between openness and secrecy. Openness is one of the most important foundations for scientific work and is essential for collaboration, peer review, transparency, reproducibility of research results, and public awareness and accountability. Openness prevents myopic thinking, dogmatism, and the spread of misinformation or disinformation. To be sure, there are times when openness is justifiably limited—to protect research subjects' confidentiality, for example—but these are rare exceptions (Resnik 2007).

Just as openness is fundamental to academic research, secrecy is core to national security. Many government programs are shrouded in secrecy and restricted by legal classification to those with the appropriate security clearances. In nondemocratic countries, such secrecy is used to shield rulers or regimes from the scrutiny of citizens. But even democratic countries classify a growing volume of data. Among that large volume, government programs and activities in the areas bearing on information security—for example, signals intelligence, lawful interception, cryptography—are among the most highly protected.

As information control has become a top national security priority, the balance between secrecy and openness has been dramatically altered in ways that affect information security research. One noteworthy example concerns the evolving nature of computer emergency response teams (CERTs). CERTs emerged in the late 1980s as institutional responses to early Internet threats, including the Morris worm. The first CERT was housed at Carnegie Mellon University, and many that followed in the United States, Canada, and Europe were also housed within universities. Not surprisingly, they shared many of the operating principles that are associated with university culture, such as transparency, peer review, and horizontal support. When a CERT at one university discovered a worm or a bug it would typically quickly transmit that information to peer CERTs to facilitate a rapid remediation of the problem. CERTS were constituent units of what was then a self-identified community of information security peers.

Over time, however, the mission and operational practices of CERTs have evolved. Many CERTs are now situated institutionally outside universities and are formally integrated into government agencies, typically within the equivalent of public safety or homeland security departments. Many of them are staffed by former or practicing intelligence or law enforcement personnel. These changes have affected not only how individual CERTs themselves function but also the community of CERTs as a whole. For example, a research report on the evolving nature of CERTs highlighted that the influence of national security agencies on CERTs has negatively affected trust and information-sharing practices (Morgus et al. 2015). CERTs are less willing to share information in ways that they did in the past, particularly concerning software vulnerabilities. While CERTs still see themselves as part of a larger network (called FIRST), their recomposition and the influence of national security pressures have degraded the CERT regime.

National Security Funding of Information Security Research

Pressures on the nature and type of information security research could also magnify as a consequence of strings tied to government-sponsored university-based research. National security funding can establish conditions that direct the sorts of issues that become worthy of study while deprioritizing others. Specifically, funding can close lines of communication across academic units or come with restrictions that inhibit open exchanges and circulation of ideas. The military and intelligence organs of the state have for many years financed and supported advanced research. For example, state funding of computing, mathematics, and physics was highly instrumental for nuclear weapons, ballistic missiles, signals intelligence, aeronautics, and space programs. Naturally, many of these programs were considered highly sensitive, and classification and secrecy were imposed on them and their researchers. In the 1960s, opposition to the Vietnam War in the United States generated controversy about the growing size and influence of classified university research projects, and there were substantial campus protests (Carlson 2007). Many academics expressed concerns that classified research would prevent scholars from freely disseminating their research findings, thus violating one of the fundamental tenets of academic freedom.

Since 9/11, the scope and scale of this type of sponsorship has grown substantially (Arkin and O'Brien 2015), while criticism has remained largely muted in the face of the now omnipresent justification for university research on the basis of creating jobs and deriving revenue. While some universities still have strict policies against classified research, many do not. Some, in fact, actively seek it out. Those that do seek it out find willing sponsors in national security agencies. In the United States, the National Security Agency (NSA) and the Department of Homeland Security (DHS) jointly sponsor the National Centers of Academic Excellence in Cyber Defense program, in which dozens of universities are enrolled.

As these programs grow in size and stature, a major concern is that they will begin to shape the nature of inquiry through a mix of positive and negative incentives. As the authors of a report on "the most militarized universities" in the United States note,

> The academy (and by extension the philanthropic world) has failed to establish a post-9/11 academic program to cultivate the next generation of scholars who can offer a genuinely civilian counter-narrative to the national security state similar

to the civilian arms control community created during the Cold War. Even at the most elite schools that rank in the top 100, the many centers and research institutes focusing on warfare and terrorism are predominantly adjuncts of the national security state. (Arkin and O'Brien 2015)

Several cases illustrate how these pressures might manifest themselves. In 2013, Matthew Green, a professor specializing in cryptography at Johns Hopkins University, wrote a blog post about how the NSA had subverted encryption standards as revealed in one of the disclosures of former NSA contractor and whistleblower Edward Snowden. University administrators formally requested that Green edit his blog post by removing links to the leaked documents and the logo of the NSA. Although they later walked back their request, some believe (Anderson 2013) the impetus for it came from an individual associated with a classified research facility on campus, the Applied Physics Laboratory, which is part of the NSA/DHS centers of excellence program.

A similar case concerns a talk that was being delivered at Purdue University by the Pulitzer Prize–winning journalist Barton Gellman (2015), who was one of the journalists for whom Edward Snowden was a source. During Gellman's public talk, he included in his slides copies of the published Snowden disclosures. Afterward, Purdue University removed the video archive of the talk and went so far as to request that the projector used during the lecture be destroyed—notwithstanding the material in question being already in the public domain. Purdue University is also a participant in the NSA/DHS program, the contract for which includes strict regulations on the handling and destruction of classified material. Gellman's experience raises a broader issue of researchers' access to leaked documents housed on controversial websites like WikiLeaks.

Some government agencies specifically ask potential employees whether they have accessed classified material without permission and will not hire those who have even if the information has been published and widely read. As an author of an article advocating the use of leaked materials for academic research notes, "Researchers may encounter significant obstacles in obtaining government grants if their research relies upon information leaked from the same government" (Michael 2015).

Another illustration concerns a case in which Carnegie Mellon University researchers deanonymized users of the anonymizer tool Tor. Their presentation at a major conference was pulled however, at the request of the

Federal Bureau of Investigation, which arrested several people identified in the research. Questions were raised as to how the bureau knew of the project in advance. An answer might lie in the university's close relationship to the NSA/DHS program. Carnegie Mellon is ranked 27th among the most "militarized universities in the United States," and the Software Engineering Institute, which is where the researchers were affiliated, "was awarded $750 million in federal funding in 2015 for both classified and unclassified work" (Arkin and O'Brien 2015; see also Hern 2016).

There could be a number of other impacts of national security funding on research. In addition to stifling free expression, national security influences can affect foreign students' participation in research. A number of US schools have signed a petition (Hern 2016) arguing that US government policies restricting foreign students from being involved in sensitive research tip the balance too far in the direction of national security over academic freedom. Research that is designated by governments as classified may increase. Historically, some types of public research, such as research on diseases and weapon systems, are classified as dual use and subject to state control. Within information security, certain types of cryptography have been the subject of classification and export controls. Research into supercomputers, quantum computing, and software vulnerabilities could also face classification. As one article notes, "Tens of thousands of patent applications are manually examined each year under the Invention Secrecy Act and referred for a final decision to the Pentagon, National Security Agency, Department of Justice and, more recently, Department of Homeland Security" (Schulz 2014). With more close involvement of national security agencies in information security research, it is reasonable to conclude that classification of research could become more common.

Just as companies can attract researchers and distort the choice of topics through positive and negative incentives, national security agencies can do the same. For example, many governments are now developing very elaborate and well-resourced psychological operations that draw from psychology, sociology, and computational techniques to predict and shape behavior. One top-secret UK Government Communications Headquarters (GCHQ) document, released as part of the Snowden disclosures, "lays out the tactics the agency uses to manipulate public opinion, its scientific and psychological research into how human thinking and behavior can be influenced, and the broad range of targets that are traditionally the province of law enforcement rather than intelligence agencies" (Greenwald and Fishman 2015). Like the NSA/DHS program, the GCHQ has sponsored cyber research centers in the United Kingdom. Cambridge University's Ross Anderson has raised concerns that programs such as these will not only steer research into topics beneficial to intelligence agencies; they will affect academic integrity. According to Anderson, "They will tell you some irrelevant and harmless top-secret fact and they will then say since this person has had access to top secret information, we need to be able to review their research for the rest of their life.... The object of the exercise is to rein in and control security research done in British universities" (Elwell 2014).

Conclusion

As I hope to have shown, what counts as a problem worth investigating in the field of information security research is not reducible to technical or functional issues, nor does it necessarily follow an objective path based on a consensus of academics. Political and economic factors can bias those decisions and affect the nature of inquiry. Approaches to information security research that challenge the status quo or highlight nontraditional forms of information security that arguably deserve greater attention can receive less support and even strong resistance.

Mitigating the trends outlined here will not be simple, nor is there one specific solution. Instead, I propose four strategic priorities to help guide efforts into broadening, diversifying, and protecting the integrity of the field.

First, interdisciplinary approaches to information security need to be encouraged toward building a more expansive discipline of information security research that includes insights from social sciences and the humanities. Although some progress has been made in this area already, more work needs to be done. To begin, we need to remind ourselves that disciplinary silos are cultural artifacts that have their basis in particular places, times, and dominant interests. This artificial and evolving character of academic inquiry is easiest to observe when academic disciplines undergo "paradigmatic change," in the words of Thomas Kuhn (1962), or are spawned by concerted efforts. To give one example, area studies was formed during the Cold War and was spearheaded by financial support from the Ford Foundation on the basis of a perceived need to educate US policy makers about distant parts of the world (Katzenstein 2002). A similar type of field-building exercise is required now around information security research. To be sure, field building is not a simple exercise. Quite apart from political, economic, and other factors outlined earlier, practical challenges abound. Disciplines speak different languages and are a lot like guilds in the sense of protecting their areas of inquiry from outside influence. This type of field building will require cross-disciplinary collaboration, which is time consuming, resource intensive, and organizationally difficult. The effort will require building up a community of peers to assist in review and other institutions that are essential to field building, such as peer-reviewed academic journals and premier conferences. In short, field building will take time, money, and extensive networking and advocacy. Fortunately, important seeds have been planted, in iSchools, in the subfield of security economics, and in the advances made in human computer interaction and user design to bridge engineering and the social sciences (see also chapter 6 by Hall, Madaan, and O'Hara).

Second, and relatedly, controversial methodologies that are essential to information security research will need extra protection. At a time when power is increasingly hard coded into the algorithms that surround us, academics need the ability to lift the lid and peer inside and beneath the surface (see chapter 1 by DeNardis about how "scholarship has to excavate and make visible the powers behind the curtain"). Reverse engineering in particular—a general term that covers a wide variety of tools, techniques, and methods—should be strongly entrenched as a core principle of academic freedom and an essential element of civic rights. Such protection may require adjustments to existing laws and regulations that ensure fair use for academics and shield them from frivolous lawsuits that aim to chill dissent. The Ontario, Canada, government's PPPA is one example. Another is Aaron's Law, proposed by the Electronic Frontier Foundation in 2013 as an amendment to the US Computer Fraud and Abuse Act in honor of the late computer researcher Aaron Swartz.

Third, to guard against the encroachment of national security and private sector influences that encourage secrecy and nondisclosure, information security researchers should strongly advocate for openness as a core principle of academic freedom: open access, open datasets, and open source code. Universities should, as a matter of core policy, reject research funding that places unreasonable limits on free speech and openness or induces other types of secrecy.

To be sure, governments will and should provide funding to universities for basic and applied research of all sorts as a matter of far-sighted public policy, and industry will do the same; but research should not come at the expense of bedrock principles that undergird the university system itself, such as transparency, reproducibility, publication of results, and peer review. Threats of classification loom large over information security research, particularly as governments move aggressively into developing information warfare capabilities. Writing in 2002, Eugene Skolnikoff noted that "the progress of science and technology and their unavoidable relevance to weapons conspire to enormously broaden the subjects that can be thought of as threats to security, and that expansion will continue long into the future." Information security researchers around the world will need to be prepared to face increasing encroachments on those aspects of their work whose open dissemination are deemed threats to the state.

Finally, universities should see information security not as a siloed and narrow technical discipline but as a core responsibility connected to the preservation of free inquiry and ultimately of knowledge itself. At a time when academic subjects are being narrowly justified on the basis of contributions to the economy, it is important to remind ourselves that the universities have their origin as institutions whose sole purpose was to nurture and protect knowledge for the betterment of the human condition. It was out of the university that many of the core principles of the Internet were born. Now that the Internet is under assault from forces of privatization and securitization, it is essential that academia recognize that it has not only a role to play but also a special obligation in protecting our public commons of information.

Notes

1. I am grateful to Masashi Crete-Nishihata, Christopher Parsons, and Adam Senft for comments on a previous draft.

2. Universities are finding it increasingly difficult to retain top faculty and students in the face of the overwhelming attraction of high-salaried job opportunities in the technology sector. See *The Economist* (2016).

References

Abbate, J. (1999). *Inventing the Internet*. Cambridge, MA: MIT Press.

Adler, E. (2008). The spread of security communities: Communities of practice, self-restraint, and NATO's post–Cold War transformation. *European Journal of International Relations, 14*(2), 195–230.

Anderson, N. (2013, September 9). Crypto prof asked to remove NSA-related blog post. *Ars Technica*. Retrieved from https://arstechnica.com/information-technology/2013/09/crypto-prof-asked-to-remove-nsa-related-blog-post/

Anderson, R. (n.d.). Economic and security resource page. Retrieved December 16, 2019, from https://www.cl.cam.ac.uk/~rja14/econsec.html

Anderson, R. (2014, May). *Privacy versus government surveillance: where network effects meet public choice* (Self-published memo). Retrieved from https://www.econinfosec.org/archive/weis2014/papers/Anderson-WEIS2014.pdf

Arkin, W. M., & O'Brien, A. (2015, November 6). The most militarized universities in America: A VICE News investigation. *VICE News*. Retrieved from https://www.vice.com/en_us/article/j59g5b/the-most-militarized-universities-in-america-a-vice-news-investigation

boyd, d., & Crawford, K. (2011, September 21). *Six provocations for big data*. Paper presented at A Decade in Internet Time: Symposium on the Dynamics of the Internet and Society. Retrieved from https://ssrn.com/abstract=1926431

Carlson, E. (2007, September 25). Scholars and secrecy—classified research comes under criticism. *NACLA*. Retrieved from https://nacla.org/article/scholars-and-secrecy-classified-research-comes-under-criticism

Citizen Lab. (2014, November). *Communities @ risk: Targeted digital threats against civil society* (Citizen Lab Report No. 48). Retrieved from https://targetedthreats.net/

Deibert, R. (2017). Cyber security. In M. D. Cavelty & T. Balzacq (Eds.), *Routledge handbook of security studies* (pp. 172–182). New York, NY: Routledge.

Deibert, R., & Crete-Nishihata, M. (2012). Global governance and the spread of cyberspace controls. *Global Governance, 18*(3), 339–361.

Elwell, M. (2014, February 14). Concern over GCHQ interference. *Varsity*. Retrieved from https://www.varsity.co.uk/news/6919

Gellman, B. (2015, October 8). Classified material in the public domain: What's a university to do? [Online forum comment]. Retrieved from https://freedom-to-tinker.com/2015/10/08/classified-material-in-the-public-domain-whats-a-university-to-do/

Greenwald, G., & Fishman, A. (2015, June 22). Controversial GCHQ unit engaged in domestic law enforcement, online propaganda, psychology research. *The Intercept*. Retrieved from https://theintercept.com/2015/06/22/controversial-gchq-unit-domestic-law-enforcement-propaganda/

Harold, S. W. (2016, August 1). The U.S.-China cyber agreement: A good first step [Blog post]. *The Rand Blog.* Retrieved from https://www.rand.org/blog/2016/08/the-us-china-cyber-agreement-a-good-first-step.html

Hern, A. (2016, February 25). US defence department funded Carnegie Mellon research to break Tor. *The Guardian.* Retrieved from https://www.theguardian.com/technology/2016/feb/25/us-defence-department-funding-carnegie-mellon-research-break-tor

Information Warfare Monitor. (2009, March 29). *Tracking GhostNet: Investigating a cyber espionage network.* Retrieved from https://citizenlab.ca/wp-content/uploads/2017/05/ghostnet.pdf

Katzenstein, P. J. (2002). Area studies, regional studies, and international relations. *Journal of East Asian Studies, 2*(1), 127–138.

Kuhn, T. (1962). *The structure of scientific revolutions.* Chicago, IL: University of Chicago Press.

Lou, E. (2015, July 30). UofT lab bane of Italian spyware firm: Leaked files. *Toronto Star.* Retrieved from https://www.thestar.com/news/gta/2015/07/30/u-of-t-lab-bane-of-italian-spyware-firm-leaked-files.html

Manovich, L. (2011). Trending: The promises and the challenges of big social data. In M. K. Gold (Ed.), *Debates in the Digital Humanities* (pp. 460–475). Minneapolis: University of Minnesota Press.

Marczak, B., Dalek, J., McKune, S., Senft, A., Scott-Railton, J., & Deibert, R. (2018, March 9). *Bad traffic: Sandvine's PacketLogic devices used to deploy government spyware in Turkey and redirect Egyptian users to affiliate ads?* (Citizen Lab Research Report no. 107). University of Toronto. Retrieved from https://citizenlab.ca/2018/03/bad-traffic-sandvines-packetlogic-devices-deploy-government-spyware-turkey-syria/

Maschmeyer, L., Deibert, R. J., & Lindsay, J. (in press). *A tale of two cyber conflicts: Civil society and commercial threat reporting.* Unpublished manuscript under review.

Michael, G. J. (2015). Who's afraid of WikiLeaks? Missed opportunities in political science research. *Review of Policy Research, 32*(2), 175–199. Retrieved from https://doi.org/10.1111/ropr.12120

Morgus, R., Skierka, I., Hohmann, M., & Maurer, T. (2015). National CSIRTs and their role in computer security incident response. *Digital Debates.org.* Retrieved from http://www.digitaldebates.org/fileadmin/media/cyber/National_CSIRTs_and_Their_Role_in_Computer_Security_Incident_Response__November_2015_--_Morgus__Skierka__Hohmann__Maurer.pdf

Pinch, T. J., & Bijker, W. E. (1993). The social construction of facts and artifacts: Or how the sociology of science and the sociology of technology might benefit each other. In W. E. Bijker, T. P. Hughes, & T. J. Pinch (Eds.), *The social construction of technological systems: New directions in the sociology and history of technology* (pp. 17–50). Cambridge, MA: MIT Press.

Resnik, D. B. (2007). Neuroethics, national security and secrecy. *The American Journal of Bioethics: AJOB, 7*(5), 14–15. Retrieved from https://doi.org/10.1080/15265 160701290264

Schulz, G. W. (2014). Government secrecy orders on patents have stifled more than 5,000 inventions. *WIRED.* Retrieved from https://www.wired.com/2013/04/gov -secrecy-orders-on-patents

Skolnikoff, E. (2002). Can traditional values survive? Massachusetts Institute of Technology Working Paper Series. Industrial Performance Center. Retrieved from https://ipc-dev.mit.edu/sites/default/files/2019-01/02-005.pdf

The Economist. (2016, April 2). Million-dollar babies. Retrieved from https://www .economist.com/business/2016/04/02/million-dollar-babies

US Department of Justice. (2014, May 19). U.S. charges five Chinese military hackers for cyber espionage against U.S. corporations and a labor organization for commercial advantage. *Justice News.* Retrieved from https://www.justice.gov/opa/pr/us -charges-five-chinese-military-hackers-cyber-espionage-against-us-corporations-and -labor

Winner, L. (1993). Upon opening the black box and finding it empty: Social constructivism and the philosophy of technology. *Science, Technology, and Human Values, 18*(3), 362–378.

Zhang, S. (2017, February 16). China's artificial-intelligence boom. *The Atlantic.* Retrieved from https://www.theatlantic.com/technology/archive/2017/02/china-artificial-intelli gence/516615

12 The Multistakeholder Concept as Narrative: A Discourse Analytical Approach

Jeanette Hofmann

Discrepancies between Ideas and Practice

Multistakeholder arrangements have a long political tradition on both the national and the international level.[1] A prominent example for the latter is the tripartite composition of the International Labor Organization, a United Nations agency founded in 1919, comprising representatives of governments, employers, and workers. The goal of tripartite organizations has been to aggregate the diversity of political positions into identifiable groups, which ideally negotiate consensual outcomes that are accepted as legitimate by all those affected, regardless of whether they directly participated in the process. Multistakeholder arrangements are a more recent variation of this model; they have emerged around cross-border or transnational issues, typically with civil society groups replacing trade unions as public interest representatives.

Over the last 20 years, multistakeholder processes have developed into a kind of new blueprint of transnational coordination. The UN Sustainable Development Goals, for example, have recognized the formation of partnerships between governments, the private sector, and civil society as a goal in itself (McKeon 2017). Likewise, the NETmundial declaration (2014) acknowledged the multistakeholder approach as the general basis of Internet governance processes. Somewhat antithetic to its rise as a role model for legitimate governance arrangements, however, is that empirical case studies have found little evidence in support of this success. Quite to the contrary, the academic literature keeps lamenting the poor performance of multistakeholder arrangements.

The apparent discrepancy between expectations and performance of multiactor approaches is itself an interesting issue to examine. A growing number of empirical studies, predominantly focusing on environmental policy, aim to understand the potential causes of the model's failures and search for ways to reduce them. Another, perhaps less obvious option to approach this discrepancy is to reflect on the model itself and the potential reasons for its rising popularity despite well-known performance problems. The second option takes an interest in the relation between concept and practices of multistakeholderism and explores this relationship from a discourse analytical perspective. It argues that discursive representations of reality are always performative; they exert a powerful impact on political processes by shaping collective perceptions of problems and their solutions, thereby giving meaning and direction to policy areas (Lynggaard 2012). For this reason, discourse analysis deserves more attention in Internet governance research.

The analysis of the multistakeholder concept touches on, and subjects to critical review, the shared knowledge that Internet governance produces about itself. As a long-term coproducer of knowledge related to Internet governance, academia should aim to include its own storylines in the analysis. No doubt, this is a difficult endeavor. One way of pursuing this goal is to focus on narratives and imaginaries as major building blocks of policy discourse. The next section introduces the concept of political narratives and imaginaries. The third section provides a short overview of the discussion on multistakeholderism from a narrative point of view. The fourth section empirically illustrates the performativity of the multistakeholder narrative, followed by a brief conclusion.

Narratives and Imaginaries

Discourses have been defined as knowledge orders consisting of "ensembles of ideas, concepts and categorizations" (Hajer 1995, 44), which ascribe meaning to the world and organize our interaction in it. A discourse can be distinguished from a mere discussion by an order that guides the creation of acceptable ideas, observations, and propositions. We thus speak of a discourse "to the extent that it is possible to register and describe a systematic set of rules for how central problems, their sources and solutions are articulated among a set of agents" (Lynggaard 2012, 90; see also Jones and McBeth 2010, 340). The analysis of public discourse can either examine actors and their discursive strategies (see Jørgensen's chapter 8) or focus on discursive artifacts and structures. This chapter is interested in the latter dimension; it studies collective meaning making as a form of narrative or plot structure relying on imaginaries and specific vocabularies such as "multistakeholderism." The common denominator of these literary constructs is the assumption that reality is always in need of representation and that any form of representation includes elements of distributed, authorless storytelling about how things really are. Even if discursive power is distributed very asymmetrically, no single actor is able to shape a public discourse. Narrativity has been characterized as a fundamental mode of "worldmaking" (Goodman 1978), and its analysis aims to decipher it as a contingent open-ended process that could always have taken a different course.

The historian and literary scholar Hayden White (1981, 2) describes narratives as a universal "metacode" that enables communicating "messages about the nature of a shared reality." Facts are selectively assembled into linear sequences that suggest a lesson. Irritating more than a few of his colleagues, White insists on the common roots of literary and political storytelling. The rhetorical strategies used, for instance, by academics to transform scattered data into an enlightening narrative, he argues, are based on the very same 19th-century plot structures as those used by novelists: satire, tragedy, comedy, and romance (White 1978). In the context of the political narrative on multistakeholderism, romances reward the struggle for the greater good by offering at least a thin silver lining on the horizon of democratic policy making.

Discourses imply narratives, and narratives, in turn, involve imaginaries or fictional elements. Fictions are not just invented; they are a necessary part of the political discourse, as Yaron Ezrahi (2012, 3–4) asserts. Well-established imaginaries such as the public sphere or civil society will help us experience fictions as facts on which rational behavior supposedly rests. Such imaginaries embody idealized representations of their subject areas and, as Charles Taylor (2004) emphasizes, they imply strong normative notions. Building on Ezrahi and Taylor, Sheila Jasanoff (2015, 4) defines imaginaries as institutionalized "collective beliefs about how society functions," how life should or should not be lived. Simultaneously, they provide the structural background against which discursive agency can evolve (Lynggaard 2012, 95).

Narratives and imaginaries constitute powerful sources of political ordering.[2] By appealing to political ideals and offering streamlined accounts of events and underlying causalities, they delimit the range of legitimate

behavior and the space of rational public discourse (Gottweis 2006). Importantly in the context of the multistakeholder concept, narratives also involve
"organizational potential" (Hajer 1995). This concerns social identities,
including classifications, strata, and roles of actors, that structurally config-
ure policy communities (for a famous example, see Anderson 1983) but also
the motivation for overcoming obstacles and realizing their mission.

Studying narratives and imaginaries implies a focus on the how of political ordering. It is less interested in the "input and output of policymaking
and their causal interrelation" (Pohle 2016, 3) than in the discursive ways
of making it work and lending meaning to it. However, there is no one best
way of conducting discourse analysis. While its origins reflect interpretative
approaches, quantitative analyses are also becoming more common (Jones
and McBeth 2010; Ten Oever, Milan, and Beraldo's chapter 10). This chapter combines a literature review on the multistakeholder model within but
also beyond Internet governance with my long-standing experience as a
participant of these processes. The next section illuminates typical accounts
of multistakeholderism to illustrate how it gains credibility and mobilizes
support amid evidence of mixed or even poor results.

The Multistakeholder Narrative: A Romantic Emplotment

As a term of art, "multistakeholder" emerged in the 1990s and gained broader
traction after the turn of the millennium. Toward the end of the 1990s,
the term began spreading across policy domains and came to also denote
private regulatory arrangements. Famous examples of the multistakeholder
approach are the Forest Stewardship Council (founded in 1993), the Global
Reporting Initiative (founded in 1997), and the World Commission on Dams
(1998–2000), the latter of which was frequently mentioned in the context
of the founding of the Internet Governance Forum (IGF). In the meantime,
multipartite bodies have also become common in areas such as global trade
and the production of consumer goods (Fransen and Kolk 2007).

The reasons for the proliferation of the multistakeholder approach have
aroused some interest in the social sciences. A common functional explanation points to coordination problems in the international sphere. According
to this view, multistakeholder arrangements are a response to the increasing number of cross-border policy issues that require cooperation beyond
the scope and competence of international organizations. The integration

of the private sector, civil society, and academia are expected to ensure the necessary degree of expertise and other resources but also compliance and support at the implementation stage.

Related explanations refer to the regulatory gaps of international policy fora, as Baumann-Pauly et al. (2017, 772) note: multistakeholder initiatives "increasingly serve a global governance function in regulating what governments leave effectively unregulated" (see also Pattberg and Widerberg 2015). Another widespread view interprets them as the result of bottom-up policy pressure. From this perspective, it is mainly civil society that is pushing for the democratization of international policy making. Giving nonstate actors a greater say in matters directly relevant for them is assumed to increase the legitimacy and effectiveness of international regulation.

Each of these explanations seems plausible. Specifically, they make sense by linking the formation of the multistakeholder approach to well-known deficits of international policy making. Multistakeholder arrangements, in other words, are presented as novel solutions to long-term structural problems of globalization. Their status as solutions confers to them a normative dimension. Seen through the lens of White's plot structure, the framing of multistakeholder efforts as a solution for intricate political problems suggests a romantic tale with a positive ending. Multistakeholder approaches seem to show that even the unruly sphere outside the nation-state can be changed for the better.

The political and, to some extent, academic discourse on multistakeholderism is characterized by storylines about how the poor state of transnational policy making can be transformed through new partnerships between various stakeholders. In particular, three recurring promises structure this narrative: the ideal of global representation, the ideal of democratizing policy making, and the ideal of improved outcomes.

Global Representation

International rulemaking has traditionally been the remit of governments and thus taken place beyond the reach of ordinary citizens. With the steady increase of transnational regulation and its impact on domestic law, nonstate actors have pointed out the lack of representation of those affected by global governance regimes. As the Cardoso report (United Nations 2004, 8) forcefully states, "The substance of politics is fast globalizing ..., the process of politics is not; its principal institutions ... remain firmly rooted at the national or local level." Citizens lack institutional means to participate in transnational policy processes and make their concerns known. With regard to Internet governance, the underrepresentation of nonstate actors seems especially problematic because the development and operation of the digital infrastructure has been predominantly private sector driven. Throughout the 2003–2005 UN World Summit on the Information Society (WSIS), the first intergovernmental process concerning itself with Internet governance, the inclusion of nonstate actors was therefore a matter of constant tension between state and nonstate actors (Epstein 2013).

The multistakeholder approach is presented as a solution to the problem of underrepresentation since it is expected to include a wide spectrum of perspectives, empower marginalized groups, and thereby form a counterforce to more powerful actors (Bécault et al. 2015). Multistakeholder processes have come to embody a redefined notion of global representation. Hence, the litmus test by which multistakeholder processes are assessed is the extent to which they manage to include the diversity of interests of those affected and to strike a power balance among them (Pattberg and Widerberg 2016).

Democratizing Policy Making

A second problem that multistakeholder initiatives are supposed to address concerns democratic deficits. Globalization weakens democracy in several ways. As Nanz and Steffek (2004, 314) observe, the concept of democratic legitimacy rests on the idea that the people set and consent to the rules that organize their political association. The decoupling of the global policy process from the constitutional apparatus of the nation-states, including the rule of law, creates a "massive democratic deficit." Moreover, traditional forms of holding political power to account do not work in global policy processes. What is needed to tackle the legitimacy deficit in the transnational sphere is new decision-making processes that incorporate principles of deliberative and participatory democracy and provide "accountability to citizens everywhere" (United Nations 2004, 24). Multistakeholder processes are assumed to achieve this goal by establishing communities of interest as a digitally enabled equivalent to territorial constituencies. They show the potential to generate new forms of procedural fairness, transparency, and accountability and thus contribute to the overdue democratization of the global sphere.

Improved Outcomes

The third challenge pertains to the overall quality of global policy making. Intergovernmental organizations are considered unable to cope with the amount, gravity, complexity, and pressing nature of today's challenges (Bäckstrand et al. 2010). Negotiation processes are found tardy, at times substantially inadequate, and leaving many policy issues unanswered owing to conflicting interests, missing expertise, dedication, and/or follow-through. A widespread disregard of human rights among governments constitutes another serious shortcoming for the area of Internet governance, which is particularly sensitive to the violation of information freedoms and privacy rights. Multistakeholder processes are regarded as a solution to these challenges because they name and shame misconduct and mobilize expertise, skills, and funding (Fransen 2012, 165). In addition to the expertise brought to the table by civil society and the private sector, it is also the learning processes enabled by a consensual style of collaboration that are said to improve the quality and legitimacy of policy outcomes (Baumann-Pauly et al. 2017; Pattberg and Widerberg 2015).

Taken together, the multistakeholder narrative exhibits a deliberative and participatory, nearly Habermasian understanding of democratic policy making with a strong emphasis on process. As Powers and Jablonski (2015, 136) nicely phrase it, this notion presumes "that strategic actors, in the right setting and by embracing shared norms, can disregard their political motivations and pressures to deliberate, listen, adjust perspectives, and come into an agreement regarding a matter of public concern." But can this assumption be regarded as an adequate description of multistakeholder policy making? Despite its increasing popularity, the overall results of multistakeholder initiatives in the transnational sphere turn out to be rather sobering.

The majority of empirical case studies report disappointing outcomes.[3] For instance, a survey by Pattberg and Widerberg (2016) on tripartite partnerships in sustainable development found that most initiatives fail not only to develop new global regulatory norms but also to improve the implementation of existing regulation or to substantially increase the integration of marginalized groups. Yet as Powers and Jablonski (2015, 152–153) observe, the significance of stakeholder inclusion for the legitimacy of policy initiatives leads to strong pressure on actors to participate, thereby considerably narrowing the room for independent criticism of the outcomes or the lack thereof. Tripartite partnerships, Pattberg and Widerberg (2016, 45) conclude, are "not just neutral instruments" for realizing agreed policy tasks; they are "sites of contestation over distinct technologies and practices." Hence, multistakeholder initiatives have their own shortcomings, and they may fail where multilateral processes have previously gone awry.

Given these empirical findings, the plausibility of the multistakeholder narrative seems to rest less on its practical achievements than on its coherence and plot. What lends credibility to the narrative are the undeniable maladies of global regulation and how it connects these to the worthwhile goals of multistakeholderism. These goals, in turn, derive their power from an imaginary that reaches beyond its immediate context of application. The striking popularity of the multistakeholder approach also originates in its reference to a metanarrative. The great promise of this metanarrative is that by implementing principles such as inclusiveness, transparency, equality, and procedural fairness, the national concept of democracy can be extended beyond territorial borders and thereby confer to transnational policy making the legitimacy it still lacks. The idea of democratizing global regulation is so powerful and uplifting that it seems to withstand all evidence to the contrary.

Yet discourse analysis is less interested in adjusting narratives or reforming malfunctioning processes than in understanding how the two worlds of narrativity and regulatory practice interact. Specifically, discourse analysis studies how narratives, once they have reached a certain degree of normality and inevitability (Taylor 2004, 17) become an enabling source of shared goals and norms, how they direct action and create a common sense of legitimacy. The next section demonstrates the stakeholder narrative at work by introducing the IGF and the Internet Corporation for Assigned Names and Numbers (ICANN) as two examples, chosen to illustrate the performative effect of the three promises described earlier.

The Multistakeholder Narrative at Work

The term "multistakeholder" entered the Internet governance landscape in 2005 during WSIS, which found that existing governance mechanisms did not provide the conditions for a meaningful participation of all stakeholder groups. The multistakeholder concept gained support for offering a middle ground between the contested alternatives of private versus public regulation of the Internet, which had paralyzed large parts of the WSIS negotiations (Musiani and Pohle 2014, 4). Following WSIS, the multistakeholder concept rapidly turned into a self-evident norm of the discourse in Internet governance. Today, it denotes a broad range of organizational models and processes (Raymond and DeNardis 2015, 14).

The epitome of the multistakeholder approach in the digital domain is the IGF. Mandated by WSIS and founded in 2006, the IGF constitutes a global space for multistakeholder policy dialogue. It is an annual conference embedded in preparatory meetings, intersessional activities, and a growing number of national and regional offspring. The second example, ICANN, is a US-based nonprofit corporation tasked with regulating the domain name system (DNS) of the Internet. It was founded in 1998, following a white paper issued by the US Department of Commerce (1998) on the premise that DNS policies should be developed by a private governance model independent of government control. Unlike the IGF, ICANN produces concrete outcomes in the form of binding policies. The present mission and legitimacy of ICANN and the IGF are firmly rooted in the imaginaries of the multistakeholder approach. Both organizations are judged by the credibility of their claims of global representativeness, their democratic standards, and their quality of output.

IGF: "Enact" the Stakeholder Taxonomy

The IGF is the first organization in Internet governance whose founding was explicitly based on the multistakeholder principle. The outcome document of WSIS (2005) stated that the IGF should "build on the complementarity between all stakeholders involved," and it named them in line with the categories used throughout the WSIS process: "governments, business entities, civil society and intergovernmental organizations." Although this classification appears rather clear-cut and simple, the stakeholders expressed from the outset uneasiness and dissent about its attributions. Civil society and the technical sector, for example, criticized the UN stakeholder taxonomy for misrepresenting them and asked for separate categories. These categories matter to the stakeholders because they determine their share of seats in committees and on workshop panels, and they also shape identities in the public discourse. The stakeholder roles, divisions, privileges, and boundaries are a permanent issue in Internet governance.

Since the geographic and political diversity of the actors involved is expanding, the stakeholder concept also struggles with problems of internal coherence. This seems mainly a problem for civil society and governments, both of which are grappling with a broad range of opinions cutting across the formal classification scheme. Recalling the enormous effort of making the stakeholder taxonomy work within the IGF, Mueller (2010, 114) notes how the "simple act" of assembling people from various sectors "for non-binding dialogue about policy can be intensely political." The stakeholders spent "countless hours" on holding the stakeholder groups together and negotiating the boundaries between them, a struggle that Mueller characterizes as "politics of representation" (Mueller 2010, 114–116).

The difficult match between the stakeholder taxonomy and the political spectrum in Internet governance is clearly at odds with the basic idea of multistakeholderism, which assumes that political positions can be aggregated along the lines of formal affiliations. Ironically, civil society, the most ardent advocate of multistakeholder representation, faces the biggest challenge in aligning its diverse membership on this model. The case of the IGF demonstrates that the stakeholder model does not constitute a natural representation of global perspectives. On the contrary, it needs to be constantly "enacted" (Epstein 2013), and a significant part of multistakeholder collaboration in the IGF is devoted to doing justice to the democratic imaginary of global representation through the never-ending re-creation of the stakeholder scheme.

ICANN: Catching Up with Democratic Standards

To be fair, the ICANN community itself never uses the term "democracy." However, the white paper from the Department of Commerce (1998) specified a set of prerequisites for the development of "sound, fair and widely accepted policies," which do qualify as democratic procedures.[4] Among them are the requirements of representation and openness and transparency and, most importantly, that the new corporation should "operate for the benefit of the Internet community as a whole" (1998, 31749). Meanwhile, ICANN (2013, 2) has updated the white paper's language and added "equality" to the criteria DNS regulation is supposed to implement: "At the heart of ICANN's policy-making is what is called a 'multistakeholder model.' This decentralized governance model places individuals, industry, non-commercial interests and government on an equal level."

ICANN's policy development process has become increasingly transparent and open to participation over the years; however, the actual decision-making authority has remained very resistant to change. The most formalized Consensus Policy Development Process in ICANN consists of no fewer than 15 steps, beginning with identifying an issue, followed by circles of reports and public comment periods, finally resulting in a recommendation to the board. No matter how inclusive, open, and fair the policy development process, the final decisions are taken by the board—on the basis of advice provided by ICANN staff, a very influential but informal filter between the bottom-up policy process and the board. While the multistakeholder narrative is driven by the idea of democratizing the transnational sphere, ICANN is still struggling to catch up with basic standards of democratic nation states. Yet remarkably, even striking democratic deficiencies are no reason for ICANN's stakeholders to question the validity of the multistakeholder approach per se. Holding ICANN's authority to account is, rather, something to be fought out, as show the intense negotiations surrounding the "Empowered Community mechanism," created in 2019. Multistakeholderism in ICANN, it seems, is less a ready-made solution for the pressing shortcomings of global regulation than a long-term agenda in itself.

IGF: Negotiating the Meaning of Outcomes

Its mandate requests that the IGF discuss, facilitate, identify, or advise on "public policy issues related to key elements of Internet governance" (WSIS 2005). How this task should be approached has been a contested issue from the start. The IGF's initial focus on enabling discourse and collaboration rather than specific recommendations was met with skepticism by those who did not believe in the impact of multistakeholder dialogue (Mueller 2015). A UN working group on "improvements to the IGF" also recommended more tangible outcomes—for example, by addressing specific policy questions and documenting the range of opinions on it (UN General Assembly 2012, 4). The IGF's cautious efforts to strengthen its efficacy notwithstanding, the actual significance of its policy dialogue is hard to determine. For some, the policy dialogue is a waste of time; for others it facilitates converging expertise, norms, and values. Epstein (2013, 147) suggests a "normalizing" role of the IGF for including nonstate actors in multilateral processes, while Mueller (2010, 122) speculates that the IGF could institutionalize our "recognition that authority over Internet governance is highly distributed."

Strikingly, the outcome of the multistakeholder dialogue is primarily assessed "through the lens of process" (Epstein 2013, 147), rather than against the background of its mandate or the many transnational policy issues awaiting attention. So far, no study has been carried out to empirically assess the impact and quality of the multistakeholder body. The proponents' focus on process supports the proposition that multistakeholderism "is sometimes viewed as a value in itself" rather than an effective form of global regulation (Raymond and DeNardis 2015, 39). In light of the multistakeholder narrative, which promises better policy outcomes, it is also interesting to note that the IGF stakeholders do not agree on what actually constitutes outcomes or on the type of outcome the IGF should strive for. The diversity of multistakeholder input thus appears as a double-edged sword; it legitimizes policy outcomes and, simultaneously, constitutes an obstacle to achieving better ones.

Conclusion: Disenchanting the Multistakeholder Narrative

This chapter starts from the premise that the discourse on Internet governance can be studied as a powerful source of political ordering. It claims that core concepts such as multistakeholderism, which are referred to by practitioners and academics alike, represent reality in a meaning-making and performative way (see Musiani's chapter 4). Narratives and imaginaries exhibit a strong normative and an organizing capacity, which influence how we interpret and engage with the world: multistakeholderism provides a sense of identity and belonging to a geographically scattered community, offers a taxonomy for defining the relationships among its members, and situates this community in a broader ideological context of competing modes of transnational regulation.

The actual achievements of the multistakeholder concept are likely to be primarily of a sensemaking nature. Its credibility is based on political aspirations rather than on a proven superior regulatory efficacy. From an empirical perspective, it is striking how much effort it takes to make multistakeholderism work. It is no exaggeration to say that the stakeholders take pains to adapt the reality of Internet governance to meet the concept's assumptions. In practice, the people involved do not easily fit into the stakeholder categories. Likewise, formal and informal authorities do not like to be held accountable, and bottom-up consensus proves to be as contested as other modes of decision-making. Multistakeholderism, it turns out, is less a regulatory approach than an end in itself; an end that shifts attention to process and requests a high degree of belief and loyalty from its followers.

What are the consequences of these findings? Is it time to abandon the multistakeholder approach? If narrativity and imaginaries are indeed a necessary, irreducible part of public discourse, as Ezrahi (2012), Stone (1997), and White (1981) suggest, the discrepancy between political aspirations and practical experiences should come as no surprise. Narratives and imaginaries may gradually fade out but not without other ones taking their place if only to enable meaningful political action. Given that narrativity cannot be skirted, Internet governance research should include discourse analytical approaches in its methodological repertoire. Moreover, it should devote systematic attention to the "worldmaking" implications of discourse (its own contributions included) and seek to dismantle its power by means of a deromanticizing critique of Internet governance narratives.

Notes

1. This chapter is an updated and shortened version of Hofmann (2016).

2. In chapter 8, Jørgensen speaks of "discursive imperialism" to denote the expanding scope of successful discourses.

3. For a more extended literature review on the reasons for the poor performance, see Hofmann (2016, 33–35).

4. Democracy can be defined in different ways. In *Democracy and Its Critics*, Robert Dahl (1989, 37) suggests five standards, which are applicable to ICANN because he intended them for associations, not for territorial nation-states. These standards are effective participation, voting equality, enlightened understanding, control of the agenda, and inclusion of adults.

References

Anderson, B. (1983). *Imagined communities: Reflections on the origin and spread of nationalism*. London, UK: Verso.

Bäckstrand, K., Khan, J., Kronsell, A., & Lövbrand, E. (Eds.). (2010). *Environmental politics and deliberative democracy: Examining the promise of new modes of governance*. Cheltenham, UK: Edward Elgar.

Baumann-Pauly, D., Nolan, J., van Heerden, A., & Samway, M. (2017). Industry-specific multi-stakeholder initiatives that govern corporate human rights standards: Legitimacy assessments of the Fair Labor Association and the Global Network Initiative. *Journal of Business Ethics, 173*, 771–787. doi:10.1007/s10551-016-3076-z

Bécault, E., Braeckman A., Lievens, M., & Wouters, J. (Eds.). (2015). *Global governance and democracy: A multidisciplinary analysis*. Leuven Global Governance Series. Cheltenham, UK: Edward Elgar.

Dahl, R. A. (1989). *Democracy and its critics*. New Haven, CT: Yale University Press.

Department of Commerce. (1998). Management of internet names and addresses: Statement of policy. Fed. Reg. 63 (111), 31741–31751.

Epstein, D. (2013). The making of institutions of information governance: The case of the Internet Governance Forum. *Journal of Information Technology, 28*(2), 137–149. doi:10.1057/jit.2013.8

Ezrahi, Y. (2012). *Imagined democracies: Necessary political fictions*. Cambridge, UK: Cambridge University Press.

Fransen, L. (2012). Multi-stakeholder governance and voluntary programme interactions: Legitimation politics in the institutional design of corporate social responsibility. *Socio-Economic Review, 10*(1), 163–192. doi:10.1093/ser/mwr029

Fransen, L., & Kolk, A. (2007). Global rule-setting for business: A critical analysis of multistakeholder standards. *Organization, 14*(5), 667–684. doi:10.1177/1350508407080305

Goodman, N. (1978). *Ways of worldmaking*. Indianapolis, IN: Hackett.

Gottweis, H. (2006). Argumentative policy analysis. In G. Peters & J. Pierre (Eds.), *Handbook of public policy* (pp. 461–479). London, UK: Sage.

Hajer, M. (1995). *The politics of environmental discourse*. Oxford, UK: Oxford University Press.

Hofmann, J. (2016). Multi-stakeholderism in Internet governance: Putting a fiction into practice. *Journal of Cyber Policy, 1*(1), 29–49. doi:10.1080/23738871.2016.1158303

Internet Corporation for Assigned Names and Numbers (ICANN). (2013, November 8). *Beginner's guide to participating in ICANN*. Retrieved from https://www.icann.org/resources/files/participating-2013-11-08-en

Jasanoff, S. (2015). Future imperfect: Science, technology, and the imaginations of modernity. In S. Jasanoff & S. Kim (Eds.), *Dreamscapes of modernity: Sociotechnical imaginaries and the fabrication of power* (pp. 1–33). Chicago, IL: University of Chicago Press.

Jones, M. D., & McBeth, M. K. (2010). A narrative policy framework: Clear enough to be wrong? *Policy Studies Journal, 38*(2), 329–353. doi:10.1111/j.1541-0072.2010.00364.x

Lynggaard, K. (2012). Discursive institutional analytical strategies: Ecological modernization and the policy process. In T. Exadaktylos & C. M. Radaelli (Eds.), *Research design in European studies* (pp. 85–104). London, UK: Palgrave Macmillan.

McKeon, N. (2017). Are equity and sustainability a likely outcome when foxes and chickens share the same coop? Critiquing the concept of multistakeholder governance of food security. *Globalizations, 17*(3), 379–398. doi:10.1080/14747731.2017.1286168

Mueller, M. L. (2010). *Networks and states: The global politics of Internet governance.* Cambridge, MA: MIT Press.

Mueller, M. L. (2015, November 16). IGF 2015: Running in place. Internet Governance Project. Retrieved from http://www.internetgovernance.org/2015/11/16/igf-2015-running-in-place/

Musiani, F., & Pohle, J. (2014). NETmundial: Only a landmark event if "digital cold war" rhetoric abandoned. *Internet Policy Review, 3*(1), 1–9. doi:10.14763/2014.1.251

Nanz, P., & Steffek, J. (2004). Global governance, participation and the public sphere. *Government and Opposition, 39*(2), 314–335. doi:10.1111/j.1477-7053.2004.00125.x

NETmundial. (2014, April). NETmundial multistakeholder statement. Retrieved from http://netmundial.br/wp-content/uploads/2014/04/NETmundial-Multistakeholder-Document.pdf

Pattberg, P., & Widerberg, O. (2015). Theorising global environmental governance: Key findings and future questions. *Millennium: Journal of International Studies, 43*(2), 684–705. doi:10.1177/0305829814561773

Pattberg, P., & Widerberg, O. (2016). Transnational multistakeholder partnerships for sustainable development: Conditions for success. *Ambio, 45*(1), 42–51. doi:10.1007/s13280-015-0684-2

Pohle, J. (2016). Multistakeholder governance processes as production sites: Enhanced cooperation "in the making." *Internet Policy Review, 5*(3) https://doi.org/10.14763/2016.3.432

Powers, S., & Jablonski M. (2015). *The real cyber war: The political economy of Internet freedom.* Urbana, IL: University of Illinois Press.

Raymond, M., & DeNardis, L. (2015). Multistakeholderism: Anatomy of an inchoate global institution. *International Theory, 7*(3), 572–616. doi:10.1017/S1752971915000081

Stone, D. A. (1997). *Policy paradox: The art of political decision making* (2nd ed.). New York, NY: W. W. Norton.

Taylor, C. (2004). *Modern social imaginaries.* Durham, NC: Duke University Press.

UN General Assembly (UNGA). (2012). *Report of the Working Group on Improvements to the Internet Governance Forum* (A/67/65–E/2012/48). New York, NY: Author.

United Nations. (2004). *We the peoples: Civil society, the United Nations and global governance*. New York, NY: Author.

White, H. (1978). *Tropics of discourse: Essays in cultural criticism*. Baltimore, MD: Johns Hopkins University Press.

White, H. (1981). The value of narrativity in the representation of reality. In W. J. T. Mitchell (Ed.), *On narrative* (pp. 1–23). Chicago, IL: University of Chicago Press.

World Summit on the Information Society (WSIS). (2005). *Tunis Agenda for the information society* (WSIS-05/TUNIS/DOC/6, rev. 1). Geneva, Switzerland: United Nations. Retrieved from http://www.itu.int/wsis/docs2/tunis/off/6rev1.html

13 Toward Future Internet Governance Research and Methods: Internet Governance Learning

Nanette S. Levinson

As this book demonstrates, there has been an increase in Internet governance research over the years, including in the number of disciplines used, the number of researchers from various countries involved, and the number of research methods implemented. Yet there are gaps (and concomitant opportunities for new work) in the methods used. This chapter provides an example of one such gap, work on policy learning as it applies to Internet governance. (Policy learning refers to "adjusting understandings and beliefs related to public policy" [Moyson, Scholten, and Weible 2017, 162].) It then poses several concluding questions regarding how we can study the broad panoply of Internet governance processes, including Internet governance policy learning in the future. It also builds on work from related research arenas to provide a foundation for these future studies, including a discussion of emerging research methods.

A Gap: Policy Learning Research and Internet Governance

To set the scene for an understanding of Internet-governance-related learning, the concept of policy space (Lambright 1976) provides an important frame. It originally referred to a single US government agency that had primary responsibility for a specific policy type. In the case of Internet governance, the US Department of Defense with its DARPA (Defense Advanced Research Projects Agency) was the original policy space for Internet studies (Braman 2011; Braman's chapter 2). This early research focusing on policy space paid little attention to the idea of learning as a process occurring in a policy space. With regard to methodology, early work on given policy spaces used the case study method and, often, stemmed from the public administration field.

Recognizing the incipient Internet's commercial and global potential, the policy space for Internet-related US governmental matters expanded over time to include other US agencies (the Department of Commerce and the Department of State), as described in Braman's chapter 2 and Mueller and Badiei's chapter 3. As the Internet expanded globally, policy spaces emerged in other national and then regional arenas and in international organization spaces. With the arrival of multistakeholderism (see DeNardis's chapter 1 and Hofmann's chapter 12), nonstate actors, including technical experts, the private sector, and civil society, began to occupy these rapidly expanding policy spaces, characterized by fuzzy boundaries.

This chapter highlights five often-interrelated types of learning that can occur in today's multiplex, nuanced, Internet-related policy space. It also calls for additional longitudinal research to capture more effectively the evolution of these processes, such as who is learning (and at what levels) over time. Note that learning can be intentional or unplanned; successful or failed. It can occur at the individual, small group, organizational, and even interorganizational levels. See figure 13.1 for an overview.

Researching fully these five types of learning calls for an examination of the processes as well as the context (characteristics of the policy spaces) and outcomes. This requires recognition of the multiple levels of analysis noted previously—from the individual to the organizational and even to the interorganizational. Adding to this complexity are related, diverse multiple

- *Governance Learning*
  - How new ideas, knowledge become acquired and used for governance purposes, especially involving information flow among state and nonstate actors
- *Network Learning*
  - How formal or informally linked sets of individuals or organizations become aware of, acquire, use, and regularize new knowledge
- *Policy Learning*
  - How individuals or organizations adjust "understandings and beliefs related to public policy" (Moyson, Scholten, and Weible 2017)

- *Interorganizational Learning*
  - How formal or informal sets of connected organizations become aware of, use, and routinize new knowledge across and within networked connections
- *Organizational Learning*
  - How "organizations identify, interpret, use, and even regularize new knowledge" (Argote and Miron-Spektor 2011)

Figure 13.1
Five learning types.

actors as the chapters in this book highlight: nation-states, regional organizations, global institutions, international organizations, civil society, and
private sector organizations. Each of these actors also evokes what I call
the culture kaleidoscope: cultural variables (including national and occupational and additional cultures and subcultures) that may (explicitly or
implicitly) shape the choice of research methods and even affect the levels
of analysis examined. Add to this the legal kaleidoscope of national laws
and regulations and regional and international laws (see Weber's chapter 5).

Lewis (2011) describes well the connection among actors in the highly
complex, networked, and even kaleidoscopic global governance policy spaces
that now exist (although her work does not focus on Internet governance).
She writes that "new forms of governance have been created to address new
governing challenges in a world where few things can be clearly separated
in meaningful ways. Network governance seems to be both the right meta-
phor to describe the increasing fragmentation, the growth of problems that
are ill-defined and which span boundaries, and the resulting dynamics of
interconnection that define contemporary governance and policy-making,
and to signal a set of governing responses to this changed environment"
(2011, 1222). While these words do not specify whether they apply to individuals or organizations in a networked policy space, they do capture well
the arena or ecosystem that is the messy global, multistakeholder, interorganizational topography of Internet governance learning today. Indeed, no
one governance policy space stands alone; thus, this chapter also calls for
the tracking of cross-policy-space learning.

The Contexts

Figure 13.2 summarizes key contextual elements (characterizing the environment or setting surrounding Internet governance or policy learning)
already identified in the preceding chapters and in the larger literature. One
element, central to a number of methodologies, is the *type of technology* and
any possible interactions among the technologies and other elements in
an environmental setting (DeNardis 2012; DeNardis's chapter 1). A second
is the *availability of resources* (munificence); resources include the availability of technology as a resource, money, people, power (Castells 2007; Nye
2011, 2014), and information supply (technical, political, social). Third is
the presence of *cultures*. I have written about the culture kaleidoscope (culture at small-group, organizational, national, diasporic, and occupational

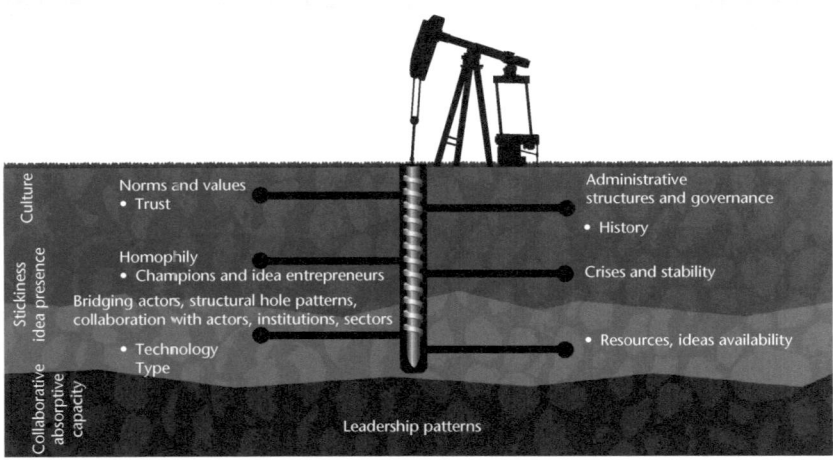

Figure 13.2
Contextual elements.

levels) (Levinson 2012). Additionally, there are age- and generation- and gender-related cultural characteristics. Each type of culture constitutes an element in the environment and has potential for shaping processes in and beyond that setting.

Fourth is the *absorptive capacity* present (Cohen and Levinthal 1990), referring to the ability of an organization to soak up new information. There are many types of absorptive capacities, ranging from the technical (the ability of an organization to absorb new technical knowledge or technologies themselves) to a general organizational learning capacity. Particularly helpful in the context studied here is the work of Easterby-Smith, Lyles, and Tsang (2008, 678) that discusses absorptive capacity with a focus on the organizational level as an "ability to recognize the value of new knowledge and to assimilate and use that knowledge."

There is, of course, an important interaction between cultures and absorptive capacity. See the work, for example, of Hofstede (1983) for a discussion of national culture and its propensity for organizational learning. Related to the notion of absorptive capacity and also cultural contexts is the work on how "sticky" new knowledge is (Szulanski 2002). As Kamkhaji and Radaelli (2017) point out in their study of European Union learning prompted by crises (another dimension of the context), the acquisition of knowledge and its utilization and routinization over time is central to policy learning.

A new dimension, yet to be analyzed in the Internet governance field, is what I call *collaborative absorptive capacity*. This refers to the ability or potential of an organization or network of organizations to collaborate whether on a joint project or policy design, implementation, or evaluation. The *degree of stability* in an environment is a fifth characteristic, followed by a sixth, the holistic *history* of a particular environmental setting. This includes trends over a long period. Here early work in the field of population ecology (Hannan and Freeman 1977) argues that a particular set of characteristics in a given environment shapes the organizations that survive over time in that environment.

Related to these characteristics is the *presence (or absence) of ideas* in an environmental setting. Diane Stone (2013, 175) puts it well when she traces, using case study and network analysis methods, how ideas created by think tanks flow transnationally. An example present in the Internet governance policy space is the concept of multistakeholderism. See the chapters by DeNardis (chapter 1), Braman (chapter 2), Mueller and Badiei (chapter 3), and Hofmann (chapter 12), which illustrate this concept, its study, and its stickiness in the Internet governance arena.

For example, the setting is rewarding (per the population ecology model) both moving away from prior governmental monopolies of telephone and telecommunications networks and moving toward coprocesses. Note that the mere presence of multistakeholderism does not imply the concomitant presence of truly collaborative processes. An additional element is the presence of *norms and values* (Finnemore and Hollis 2016; Henry 2011). This dimension of the environment is even more relevant currently, as various actors grapple with developing and articulating norms and values related to platforms.

Next, we need to pay attention to the *actors* in the context. Who or what unit is doing the learning? The presence or absence of *homophily* (perceived similarity among individuals) fosters trust in learning processes (Reagans 2011). There is also a need for clarity on which levels of analysis are studied and on any interactions among these levels (Pahl-Wostl 2009). Another view of actors outlines the roles they play. Some serve as bridging actors (Spekkink and Boons 2015), occupying roles that link organizations. They can, indeed, also bridge structural holes and foster the flow of ideas across organizational and network boundaries (Burt 2009). Note that new social media can instantaneously bridge structural holes. (With regard to bridging, boundaries and boundary organizations are important concepts in science and technology studies research (see DeNardis's chapter 1 and Musiani's chapter 4).

The term "actors" also goes beyond the individual level to include the organizational and even the institutional (Craft and Wilder 2017): governments, private sector organizations, and civil society organizations and institutions (including international organizations such as the Organization for Economic Cooperation and Development (OECD) or the UN's Educational, Scientific, and Cultural Organization (UNESCO), International Telecommunication Union, and Conference on Trade and Development (UNCTAD) that are of much interest in Internet governance and related global governance studies. Galaz et al. (2017, 11) find "complex institutional interactions and actor constellations crisscrossing institutional levels, sectoral policies, and established public-private partnerships." The flow of information across these interactions, as noted earlier, is important to track.

We also see the emergence of new institutions in policy spaces. An outcome of the UN-sponsored World Summit on the Information Society (WSIS) was the 2006 creation of an entirely new entity—distinctive, dialogue-focused, and multistakeholder in nature but not decision-making—the Internet Governance Forum (IGF) (see Braman's chapter 2, Mueller and Badiei's chapter 3, Cogburn's chapter 9, and Hofmann's chapter 12). This multistakeholder setting can be a locus for policy learning. (See Levinson and Marzouki [2016] for examples of such learning episodes.)

Finally, there are four additional and relevant characteristics of the environment. First, Weber and Khademian's (2008) work highlights both initial reputation-based *trust* and prior development of knowledge-based trust as key contextual elements for successful interorganizational transfer of practices. Second, additional work examines *governance patterns* of any formal sets of organizations or formal networks and their effects on idea flow and use. Third, the public administration field (Stone 2019) yields research that identifies the presence or absence of potential or actual *champions or idea and policy entrepreneurs* as key factors in idea or policy transfer. Fourth, research in the environmental governance field (Wurzel, Liefferink, and Tomey 2019) highlights *leadership* patterns at the individual, organizational, or even network (interorganizational) levels in terms of such impacts.

The Processes

What is taking place with regard to policy and Internet governance learning processes? To which concepts should we pay attention to understand the nuanced happenings in Internet governance processes, especially policy learning, organizational learning, or governance learning? Each of these processes has a representative literature with slight cross-fertilization. (See figure 13.3 for a list of major learning processes that can be applied to Internet governance and its ecosystem.)

Separate but sometimes overlapping bodies of research are dedicated to policy learning, organizational learning, and more recently, network learning and governance learning. With regard to organizational learning, scholarship most often originates in the business management or organization and cognitive psychology fields, primarily with private sector arena focus—again with little cross-disciplinary dialogue (see Argote and Miron-Spektor 2011). The work on policy learning and network or governance learning tends to come from the public administration and political science fields. These scholars use research methods common in their fields. Finally, while governance learning as a concept dates back to at least Ruggie's (2001)

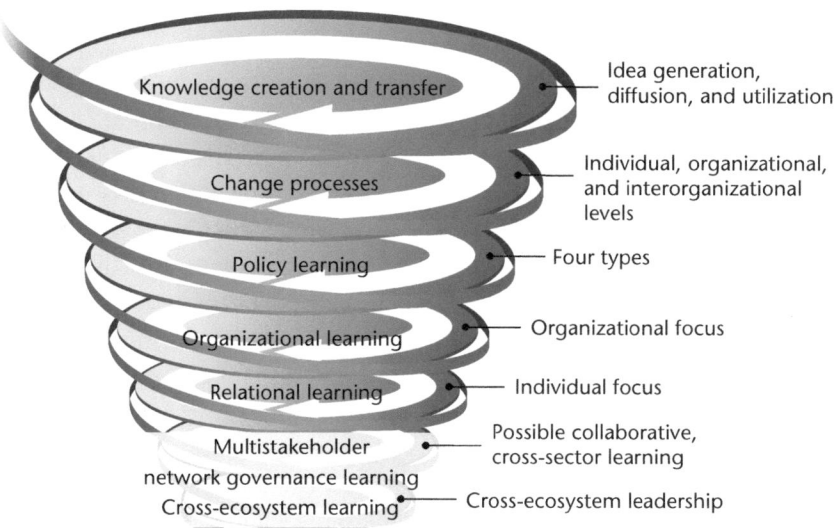

Figure 13.3
Learning processes.

work, more recent scholarship (especially in the area of environmental governance) examines network governance and network governance learning, as discussed later. Here the unit of analysis is more than one government agency and can include nonstate actors or organizations as well.

Policy Learning, Organizational Learning, and Network Governance

"In short, network governance rests on a recognition that policy is the result of governing processes that are not fully controlled by governments. Policy-making occurs through interactive forms of governing that involve many actors from different spheres" (Lewis 2011, 1222). Lewis's analysis applies well to the multistakeholder setting in much of Internet governance work.

Research on the attitudes that government policy makers bring to their work and that may shape policy learning is emerging. Jeffares and Skelcher (2008) used social network analysis and pioneered a Q methodology that allowed them to survey the attitudes of English and Dutch civil servants to understand better what these policy makers bring to the table. They identified different attitude clusters: pragmatism, realism, adaptation, and optimism. Having Internet governance research that focuses on stakeholders in multistakeholder settings and their attitudes and discourse as they begin to interact in such settings provides a valuable addition to our knowledge base. (See Cogburn's chapter 9 and Hofmann's chapter 12 for examples.)

Argote and Miron-Spektor (2011, 1124) provide an overview of organizational learning and define it as "a change in the organization that occurs as the organization acquires experience"; it involves three sub-processes: "creating, retaining and transferring knowledge." There are differences between organizational learning in the public sector and in the private sector, thus possessing implications for networked governance (with a focus on networked organizations) learning. Visser and Van der Togt (2016) highlight these sectoral differences and link them to differing scholarship traditions in public administration compared with business management.

Governance learning itself relates to "the processes and procedures of multistakeholder decision-making and planning" (Challies et al. 2017, 289). Challies and colleagues study novel European Union water governance formats that are multistakeholder and multilevel in nature; they are especially interested in how policy makers design and run participatory processes. This work can inform Internet governance studies, in which often the same participatory, multistakeholder phenomena are at work. (See also Apgar et al. 2017.)

Newig, Günther, and Pahl-Wostl (2010) and Newig et al. (2016) underline various types of governance learning. These include parallel learning (from other policy fields), serial learning, learning from endogenous sources, and learning from exogenous sources (from other jurisdictions and from other fields). Thus, the emphasis here on characteristics of contexts and the actors in those contexts is key regarding learning from exogenous sources.

The field of Internet governance needs additional studies of governance learning processes, whether developing- or developed-nation policy spaces. What are the actual knowledge flow processes across individuals and across unlike organizations over time? Is new knowledge or evidence actually used and, if so, how in governance processes? Does this new knowledge stick? What research methods work well for such studies?

Linking governance learning (which goes beyond organizational learning) to policy learning, Lewis (2011) sees policy as the result of multistakeholder-infused governing processes. Thus, she argues that governments no longer have total control of policy processes. Further, Dunlop and Radaelli (2013, 603) note that policy learning can involve four types of learning: reflexivity, pluralistic bargaining, epistemic, and hierarchical.

While Dunlop and Radaelli focus on European fiscal policy (finding the pluralistic bargaining and hierarchy modes most prevalent in their study of two European Union countries), their policy learning typology holds power for other global governance domains. Internet governance policy learning has the following additional categories, each of which can have present one or more of the Dunlop and Radaelli types: policy learning related to design, implementation, or evaluation.

A fascinating addition to the literature is work on policy learning via policy experiments. This work links policy learning to environmental governance and defines policy experimentation as "a process that generates learning through an explicit intention to test new ideas" (McFadgen and Huitema 2017, 1767). It also highlights a category of learning McFadgen and Huitema term "relational learning"—"a change in trust, the ability to cooperate, and understanding of other parties" that occurs in policy experimentation. Calling for future research to consider quality of leadership, demographics, and motivations of participants, this work possesses much potential for understanding the complex Internet policy space, especially those spaces involving multistakeholder participation. A rare example of an experimental approach in the Internet governance field is the Fishkin et al. (2018) study in which the researchers use deliberative polling as a data collection tool.

Knowledge Creation and Transfer Processes

In this complex and relatively new multidimensional policy space, knowledge creation, dissemination or transfer, and use become key targets of study, along with innovation creation, diffusion, and use. This literature has had its own research arena, primarily in the communication studies field (Rogers 1962), with little cross-fertilization until the advent of the Internet. There is much opportunity now to trace an idea or practice or policy moving from the environmental governance arena to the Internet governance and security arenas or vice versa. We also need to consider the time, or longitudinal element, in such learning.

Contemporary work in the field of policy making increasingly highlights evidence-based policy making (Head 2016). There is some research that focuses on knowledge brokers and how they might promote information transfer. Howlett, Mukherjee, and Koppenjan (2017) identify five types of brokers, or bridging actors: coordinators, consultants, gatekeepers, representatives, and liaisons. Looking at governance learning or collaborative governance learning possibilities, Head identifies eight main knowledge transfer and exchange modes for brokers (2016, 478–479):

- Face-to-face exchange between policy makers and researchers
- Educational meetings for policy makers
- Networks and communities of practice
- Meetings with facilitation between researchers and policy makers
- Cross-disciplinary and interactive workshops
- Capacity building
- Web-based information and communication
- Steering committees that interpret research and integrate local experts' ideas into design

This work also underlines the absence of effective brokering. It may be—future research will inform us—that the coprocess and participatory approach detailed earlier can minimize the need for brokering and maximize the effective incorporation of research or evidence from practice into policy making. Yet it may also be true that brokering remains an important function, given the highly technical nature of Internet infrastructure and, especially, emerging and converging Internet-related technologies.

There is a need, too, for more cross-policy-space work on what catalyzes policy learning in each domain and across global governance domains. Some of the works cited here include a focus on crises as a catalyst for policy learning, but further research is needed to understand the full panoply (as well as rich impacts) of catalytic factors. Future research should consider findings (and methods) from work on improvisation and bricolage (Crossan 1998) as responses to crises or uncertainty.

Additionally, the actual presence of informal or formal networks, as in multistakeholder contexts, gives rise to possible coprocesses leading to collaborative governance learning or absence thereof (Baird, Plummer, and Bodin 2016; Boivard et al. 2016). Moyson (2017) discusses cocreation processes in three settings and highlights how context counts in Estonia, the Netherlands, and Germany. Elsewhere, Levinson and Marzouki (2016) begin to analyze coprocesses in the context of global Internet governance, using document analysis and in-depth interviews.

While Internet governance research has begun to analyze coprocesses, it has taken little notice of research using the advocacy coalition framework (ACF) approach. That approach appears primarily in the political science and public administration literature. Weible, Sabatier, and McQueen (2009, 132) define "(advocacy) coalitions as consisting of members who share policy core beliefs and engage in a nontrivial degree of coordination." There is a need to compare this ACF approach (Weible and Carter 2017) with a multistakeholder approach (Raymond and DeNardis 2015; see also Jørgensen's chapter 8 and Hofmann's chapter 12) wherein stakeholders can and do have differing policy core beliefs. The ACF links to knowledge transfer and policy learning as a result of advocacy coalitions sharing information with policy makers, whereas a multistakeholder approach encompasses greater complexity with regard to the disparate cultures, power panoplies, sectors, countries, institutions, and individuals involved, each with distinct perspectives and policy beliefs. ACF shares research methods with Internet governance scholars. This includes the use of interviews, surveys, and document and archival analysis; ACF work also emphasizes the need for a longitudinal dimension, often calling for a 10-year period for analysis.

Beginning with either the birth of the Internet Engineering Task Force (approximately 1986), the network of technical experts from around the world who came together and continue to meet to craft many Internet-related standards, or the creation (in 1998) of the Internet Corporation for

Assigned Names and Numbers (ICANN) (Klein 2002; Kleinwächter 2000; also see DeNardis's chapter 1 and Ten Oever, Milan, and Beraldo's chapter 10), we can trace knowledge transfer episodes both within and across unlike organizations. Indeed, we can also trace the growth of multidirectional learning both within informally connected networks and across formal network organizational arrangements. Networked organizations concerned with Internet policy matters and across global governance policy domains also provide rich examples of knowledge transfer and coprocess episodes (Levinson 2015).

Elsewhere, using case study approaches and content analyses rooted in communication and organizational sociology, Levinson (2012) has documented the dissemination and use of multistakeholder approaches, which are especially evident in the IGF, created as a multistakeholder entity in 2006. The IGF's advisory board, at its inception, was called simply the advisory group. It was not until almost a year later that the name was changed to the Multistakeholder Advisory Group. Central to this change was the executive director of the IGF, Nitin Desai, who formerly had worked on environmental governance challenges at the United Nations and, thus, had become aware of the idea of multistakeholder approaches at an environmental summit before taking on his role in the Internet governance arena.

The IGF by its very constitution is not a decision-making or direct policy-making body, but it provides a locus for informal policy learning. The IGF website itself advertises that the IGF is a place designed for cross-sector or multistakeholder dialogue, thus having the potential to catalyze policy learning.[1] The annual IGF meetings, held in locations throughout the world, place proceedings' transcripts online. Content analyses of these proceedings provide evidence of the transfer of ideas through awareness raising and dialogue (see Cogburn's chapter 9); this knowledge flow can be traced to nation-state government decision-making. For example, the announcement of the US Department of Commerce decision to end much of its contractual relationship with ICANN in its historic affirmation agreement with ICANN in 2009 uses the language of multistakeholder, international bottom-up governance, echoing discourse at the IGF.

In the case of cybersecurity, the September Group of Seven Declaration in Turin, Italy, (Group of Seven 2017, declaration point #6) highlights a similar phenomenon. It calls for governments to have increased cooperation with the private sector and beyond: it is "crucial to engage proactively with the private sector, scientific community, academia, the technology community and civil society in an open, inclusive and transparent approach to developing our policy responses and initiatives." And its annex 3 on cybersecurity underlines the need for cooperating with the private sector. Trust, awareness, and information sharing emerge as important.

Carrapico and Farrand (2017) underline this increasing role of the private sector in cybersecurity and cyberspace policy. They and Van Eeten (2017, 433) find a "surprising amount" of voluntary activity. The expansion of Internet governance policy space to include private sector and even civil society actors also gives rise to accountability questions (Eggenschwiler 2017) and to power-differential questions related to learning processes. Deibert's chapter 11 adds to the accountability dimension and provides an in-depth view of power roles in Internet governance policy making. Studying these dimensions calls for research methods such as in-depth qualitative or experimental approaches that can capture rich nuance.

The Outcomes

Much more work in the area of outcomes is needed, in at least two categories: the identification of outcome types and of research methods for studying the presence or absence of collaborative learning and governance (Baird and Bodin 2016; Challies et al. 2017). See figure 13.4 for an overview. Kamkhaji and Radaelli (2017) talk about policy learning itself as an outcome, particularly in the face of crisis, and argue that policy learning follows change. Similarly, Or and Aranda-Jan (2017) write about governance after the global financial crisis and underline civil society joining nation-states as key actors.

Some public administration literature examines outcomes in terms of network effectiveness (and as a subset, network management) (Provan and Kenis 2008; Provan and Lemaire 2012; Steijn, Klijn, and Edelenbos 2011). Provan and Lemaire (2012) also highlight barriers to network effectiveness (with a focus on US public service delivery) while Raab, Mannak, and Cambré (2013) track how network structure, context, and governance relate to network effectiveness in the context of 39 interorganizational networks focused on reducing recidivism in the Netherlands. Note that Raab, Mannak, and Cambré examine outcomes at the community level. They find that network effectiveness depends on how aligned the strategy and goals, governance mode, structure, people, and management processes are and

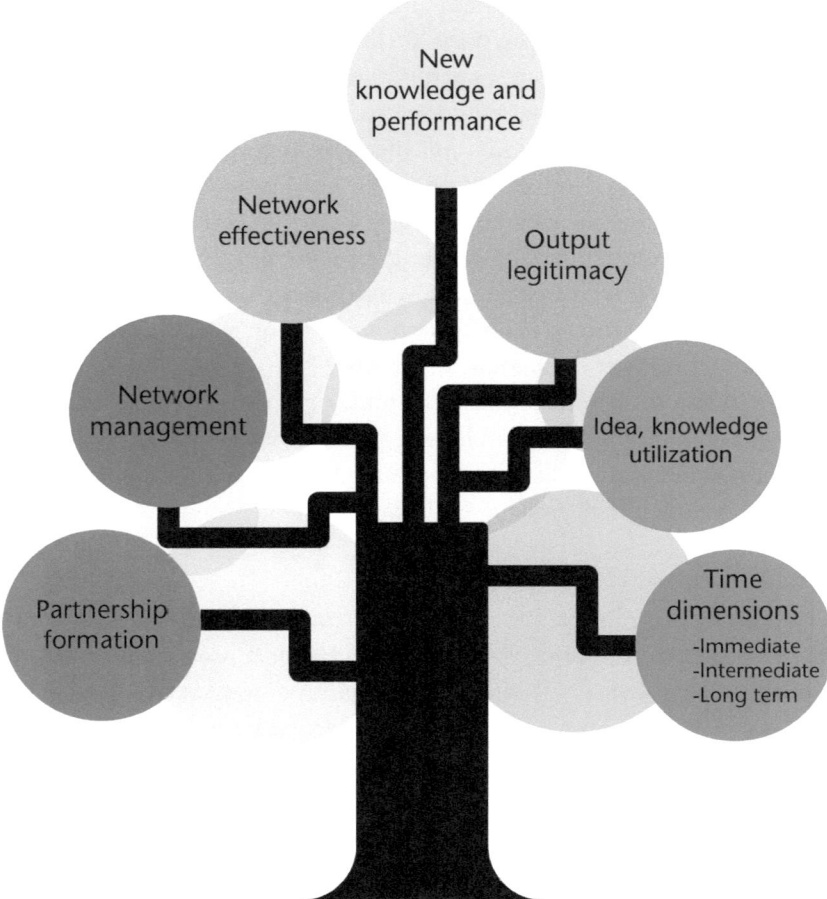

Figure 13.4
Outcomes.

how they match the environmental context. There is also work on leadership of networks and network effectiveness (Busch and Barkema 2019; Wurzel, Liefferink, and Tomey 2019).

Scholars have already studied two specific outcome types in policy learning settings: the development of new knowledge and the change in beliefs (Leach et al. 2013; Moyson 2017). Moyson (2017) examined consistency and cognition in policy learning, finding that core policy beliefs in the groups he studied appear to be stable over time.

Other research (Allan 2017, 131) focuses on climate change and global governance and describes outcomes as the result of a "dynamic, interactive process between states and scientists" rather than a one-way knowledge transfer from scientists to states. Using a science and technology studies approach, Allan illustrates the ways in which political figures play a key role in knowledge production; further, he argues that on the basis of his research one cannot separate the science (and scientists) from politics. In sum, his study provides an example of nuanced, nonlinear coproduction processes and also possesses implications for how policy makers use information. (See also Weiss [1979] for an early study of how US policy makers use information and Nwagwu and Iheanatu [2011] for a study of Nigerian policy makers' use of information.)

In the health and the security arenas, there are similar discussions regarding research on outcomes. Krahmann (2017) underlines performance measurements as outcomes and argues for adopting a performative approach that finds international, as opposed to local, actors determining outcomes. This work also raises ethical issues, as the author questions "output legitimacy" (Krahmann 2017, 60).

Another study (Galaz et al. 2017) argues that research does not give enough credence to the power and roles of international institutions and networks. The particular focus of this work is risks related to climate change, food crises, and financial crises. In sum, they find that international institutions do have impacts on state institutions. This corresponds to findings from Levinson and Marzouki (2016) in the Internet governance arena in which international organizations play complex roles with regard to generating and sustaining novel ideas that are not always consonant with individual member states' positions. These ideas feed into the policy learning process at international levels. Finally, work on land mines discusses the power of nongovernmental organizations and networks to bring about policy change through nation-state participation in an international setting (Rutherford 2000).

Research Methods and Approaches

The work on international institutions and networks and the complexity and turbulence that characterize the Internet governance field raise a significant research challenge similar to that posed in DeNardis's chapter 1: How can we best study the outcomes of learning processes related to coproduction (or other), whether the coproducing involves the effectiveness (or management) of a network or the design or implementation or even evaluation of a policy or simply governance learning over time? This book's chapters provide a rich and representative portfolio of methods, including qualitative, quantitative, and mixed methods. These include archival analysis, discourse analysis, automated text analysis, social network analysis, interviews, focus groups, participant observation, surveys (including online surveys), and case studies. Experiments and simulation also supply rich and nuanced data.

As we consider measuring or analyzing outcomes, it is vital to incorporate a longitudinal time frame (Moyson 2017; and see Jardine's chapter 7). As change logic analysts argue, there are at least three time-linked dimensions in outcome categories: outputs (immediate), outcomes (intermediate time range), and impacts (long-term time range). See, for example, the Kellogg Foundation's Logic Model Development Guide (2004).

Moreover, there are many organizational learning contexts on which to focus, each of which has its own culture and absorptive capacity. For example, the United Nations, its UNESCO, and its International Telecommunication Union subsystems are real examples of the presence or absence of interorganizational and cross-sector learning. Or one can use ICANN or the Internet Engineering Task Force or IGF as focal organizations, as do many of the chapters in this book. One can also explore intraorganizational or endogenous learning; in the UN system, for example, one can study staff roles vis-à-vis member-state representative roles within each organizational unit.

Researchers who work on specific global governance arenas have an important opportunity now to undertake additional comparative studies that include attitudinal dimensions and examine organizational learning, policy learning, and networked governance learning contexts, processes, and especially, outcomes. Research methods need to go beyond those discussed here and be shared across governance arenas. As technologies emerge and converge, there are numerous opportunities to innovate and experiment in our studies.

While this chapter provides the example of policy experiments as a type of policy learning, there is much potential also for actual experiments to illuminate organizational, policy, and governance learning. There is also an important opportunity and need to examine intricate interrelationships among organizational, policy, and governance learning and to, especially, incorporate cultures and infrastructure (Musiani et al. 2016; see also DeNardis's chapter 1 and Musiani's chapter 4) into the discussion. To do this, a combination of qualitative and quantitative methods is necessary. One approach discussed earlier, that of science and technology studies scholars (DeNardis's chapter 1; Musiani's chapter 4), possesses much potential (see also Epstein, Katzenbach, and Musiani 2016; Katzenbach et al. 2015). Another approach used in global governance policy spaces is that of participatory action research. This participatory focus is particularly powerful in analyzing policy spaces characterized by collaborative or coprocesses.

Another new approach focuses on the individual level and incorporates methods from cognitive psychology to capture changing belief processes and learning. Finally, the time is ripe for using knowledge transfer theory and methods regarding any cross-governance (and cross-cultural) policy space learning, especially with a focus on absorptive capacity.

Shared Research Challenges: Implications for Future Studies

Internet governance (including policy learning) shares at least six research challenges with other global governance arenas such as environmental or health governance:

- A technical or scientific dimension
- Multiple levels of analysis (from the individual to the interorganizational)
- Increasing involvement of nonstate actors
- Multilayered interactions including cross-border, cross-cultural, and transnational
- Complex regulatory questions
- Uncertainties (including technological developments and convergences) and turbulence that contribute to added complexity

Taken together, the chapters in this book also highlight what is distinctive about the Internet governance field we research. The five elements below, from a research perspective, echo the five features that DeNardis in chapter 1, using a practice perspective, argues distinguish Internet governance:

- The nature of the Internet itself—the infrastructure
- The inextricable links among infrastructure and social, political, and economic dimensions

- The time dimension with the Internet's instantaneous impacts
- The information intensities of Internet-related activities
- The emerging synergies with science (e.g., nanotechnologies), mechanics (e.g., the Internet of Things), and other yet undetermined fields

The conceptual frames and research methods we use need to match these complex and dynamic characteristics. Clearly, as illustrated in this book, there is no one method to capture all the complexities of Internet governance as it exists in 2020 and in the years ahead. Yet the methods presented here do presage emerging methods, including the increasing use of ethical experiments and clinical models for the conduct of research. Cross-research field dialogue such as can be seen in Hofmann's chapter 12 with its discussion of the environmental global governance field can facilitate sharing methods that work and do not work in one's own research arena. This can be especially useful with regard to the "landscape of tensions" (Gustafsson and Lidskog 2018, 4) among multilayered governance, multiple levels of analysis, and cultural complications that exist in many global governance fields. Such dialogue can assist Internet governance researchers (and especially civil society research consumers) to get out ahead on complex policy issues. (Some have argued that industry, especially in the cybersecurity arena, is out in front of the rest, e.g., governments and civil society.)

Emerging Research Methods and Opportunities Ahead

Research methods developed for studying cross-cultural communication at the individual, team, and organizational levels can be useful in recognizing intercultural interactions in multistakeholder settings and related learning processes. Here there is scant research work on stakeholder, cross-cultural dialogue styles and stakeholder power-related strategies. Nor is there much work with a focus on gender. To measure these dimensions, studies outside the Internet governance field have used self-reported survey measures (Schneider and Heinecke 2019). Possible methods for analyzing these dimensions for the Internet governance field are simulations, games, and experiments. Another underexplored research arena is that of indigenous research methods. Such methods (Schneider and Heinecke 2019) could be powerful in studying collaborative processes such as those in which local communities are a part of codesign, coimplementation, or coevaluation.

Internet governance research focuses on the emergent topic of platform governance (DeNardis and Hackl 2015; Gorwa 2019; Hein et al. 2016; Schreieck et al. 2018). It primarily serves to frame, define, and raise important questions with regard to private sector technology-based platforms, including social media platforms. Little attention has yet been paid to the methods we use for researching platform governance. See Jørgensen's chapter 8 for one useful example. Additionally, there are opportunities to study new blockchain technologies as governance platforms. Here there are exciting questions about how best to research blockchains as platform governance, including examinations of power within the blockchain.

There is new and relevant work in the public administration field on governance platforms (Ansell and Gash 2018; Ojo and Mellouli 2018). Unlike platform governance research, which focuses on actual private sector technology platforms (excluding blockchain, in which involvement goes beyond the private sector), the public-administration-based research on governance platforms focuses on nongovernment as well as government involvement in governance networks. Ansell and Gash (2018, 19, 30) coined the term "governance platform," inspired by the presence of platform governance. Future work on platform governance, governance platforms, and related research methods could also add to our understanding of norm development, another area of emerging importance in Internet governance locally and globally. Indeed, a dialogue between platform governance and governance platform researchers could advance both research arena agendas.

As we study the path of Internet governance locally and globally, a question of long-term importance arises: How do societies develop what some have called dynamic resilience (Ansell and Trondal 2018; Fonseca, Lukosch, and Brazier 2019)? How do we study this? Certainly, we need a longitudinal dimension for such a study. Crises do and will emerge over time. (See research on Kondratieff curves and international affairs in Goldstein [1991].) What research methods, perhaps in combination with economic analysis, might best fit this research question as it applies to the complex and dynamic field of Internet governance studies? On the basis of the work of Ansell and Trondal (2018), researchers can consider the following approaches: observation, improvisation, conduct of experiments, tracking of networks, and tracing other interinstitutional arrangements through interviews, observation, and network analysis. Could researchers craft an index of societal dynamic resilience with regard to Internet governance? A focus on dynamic resilience could provide rich data, given complex emerging technologies such as artificial intelligence and robotics.

A concluding question relates to the funding of research and research methods. How does the pattern of funding for research questions (and research methodologies) shape, if at all, the portfolio of research frames (including disciplines and cross-disciplines) and approaches to be applied in the work ahead? Stone (2013) highlights the ways in which think tanks shape idea diffusion. What might be the road ahead for future funders of Internet governance research? And always mindful of ethics (see especially Hall, Madaan, and O'Hara's chapter 6; Jardine's chapter 7; and Ten Oever, Milan, and Beraldo's chapter 10) and power (see especially Deibert's chapter 11), where might that road lead in terms of Internet governance research parameters, methods, and outcomes?

Note

1. Internet Governance Forum, accessed December 21, 2019, https://www.intgov forum.org/multilingual/.

References

Allan, B. B. (2017). Producing the climate: States, scientists, and the constitution of global governance objects. *International Organization, 71*(1), 131–162.

Ansell, C., & Gash, A. (2018). Collaborative platforms as a governance strategy. *Journal of Public Administration Research and Theory, 28*(1), 16–32.

Ansell, C., & Trondal, J. (2018). Governing turbulence: An organizational-institutional agenda. *Perspectives on Public Management and Governance, 1*(1), 43–57.

Apgar, J. M., Cohen, P. J., Ratner, B. D., deSilva, S., Buisson, M.-C., Longley, C., Bastakott, R. C., & Maedza, E. (2017). Identifying opportunities to improve governance of aquatic agricultural systems through participatory action research. *Ecology and Society, 22*(1), 9.

Argote, L., & Miron-Spektor, E. (2011). Organizational learning: From experience to knowledge. *Organization Science, 22*(5), 1123–1137.

Baird, J., Plummer, R., & Bodin, O. (2016). Collaborative governance for climate change adaptation in Canada: Experimenting with adaptive co-management. *Regional Environmental Change, 16*(3), 747–758.

Bovaird, T., Stoker, G., Jones, T., Loeffler, E., & Pinilla Roncancio, M. (2016). Activating collective co-production of public services: Influencing citizens to participate in complex governance mechanisms in the UK. *International Review of Administrative Sciences, 82*(1), 47–68.

Braman, S. (2011). The framing years: Policy fundamentals in the Internet design process, 1969–1979. *The Information Society, 2*, 295–310.

Burt, R. S. (2009). *Structural holes: The social structure of competition.* Cambridge, MA: Harvard University Press.

Busch, C., & Barkema, H. (2019). Social entrepreneurs as network orchestrators. In G. Gerard, T. Baker, P. Tracey, & H. Joshi, *Handbook of inclusive innovation* (pp. 464–484). Cheltenham, UK: Edward Elgar.

Carrapico, H., & Farrand, B. (2017). Dialogue, partnership and empowerment for network and information security: The changing role of the private sector from objects of regulation to regulation shapers. *Crime Law and Social Change, 67*, 245–263.

Castells, M. (2007). Communication, power and counter-power in the network society. *International Journal of Communication, 1*, 238–266.

Challies, E., Newig, J., Kochskämper, E., & Jager, N. W. (2017). Governance change and governance learning in Europe: Stakeholder participation in environmental policy implementation. *Policy and Society, 36*(2), 288–303.

Cohen, W. M., & Levinthal, D. A. (1990). Absorptive capacity: A new perspective on learning and innovation. *Administrative Science Quarterly, 35*(1), 128–152.

Craft, J., and Wilder, M. (2017). Catching a second wave: Context and compatibility in advisory system dynamics. *Policy Studies Journal, 45*(1), 215–239.

Crossan, M. M. (1998). Improvisation in action. *Organization Science, 9*(5), 593–599.

DeNardis, L. (2012). Hidden levers of Internet control. *Information, Communication & Society, 15*(5), 720–738.

DeNardis, L., & Hackl, A. M. (2015). Internet governance by social media platforms. *Telecommunications Policy, 39*(9), 761–770.

Dunlop, C. A., & Radaelli, C. M. (2013). Systematising policy learning: From monolith to dimensions. *Political Studies, 61*(3), 599–619.

Easterby-Smith, M., Lyles, M. A., & Tsang, E. W. K. (2008). Inter-organizational knowledge transfer: Current themes and future prospects. *Journal of Management Studies, 45*(4), 677–690.

Eggenschwiler, J. (2017). Accountability challenges confronting cyberspace governance. *Internet Policy Review, 6*(3), 1–11.

Epstein, D., Katzenbach, C., & Musiani, F. (2016). Doing Internet governance: How science and technology studies inform the study of Internet governance. *Internet Policy Review, 5*(3), 3–14.

Finnemore, M., & Hollis, D. (2016). Constructing norms for global cybersecurity. *American Journal of International Law, 110*(3), 425–479.

Fishkin, J. S., Senges, M., Donahoe, E., Diamond, L., & Siu, A. (2018). Deliberative polling for multistakeholder Internet governance: Considered judgments on access for the next billion. *Information, Communication & Society, 21*(11), 1541–1554.

Fonseca, X., Lukosch, S., & Brazier, F. (2019). Social cohesion revisited: A new definition and how to characterize it. *Innovation: The European Journal of Social Science Research, 32*(2), 231–253.

Galaz, V., Tallberg, J., Boin, A., Ituarte-Lima, C., Hey, E., Olsson, P., & Westley, F. (2017). Global governance dimensions of globally networked risks: The state of the art in social science research. *Risk, Hazards & Crisis in Public Policy, 8*, 4–27.

Goldstein, J. (1991). The possibility of cycles in international relations. *International Studies Quarterly, 35*(4), 477–480.

Gorwa, R. (2019). What is platform governance? *Information, Communication & Society, 22*(6), 854–871.

Group of Seven. (2017, September 25–26). *G7 ICT and Industry Ministers' declaration: Making the next production revolution inclusive, open and secure.* Turin, Italy. Retrieved from http://www.g7italy.it/sites/default/files/documents/G7%20ICT_Industry_Min isters_Declaration_%20Italy-26%20Sept_2017final_0/index.pdf

Gustafsson, K. M., & Lidskog, R. (2018). Boundary organizations and environmental governance: Performance, institutional design and conceptual development. *Climate Risk Management, 19*, 2–11.

Hannan, M. T., & Freeman, J. (1977). The population ecology of organizations. *American Journal of Sociology, 82*(5), 929–964.

Head, B. W. (2016). Toward more evidence-informed policy making? *Public Administration Review, 76*(3), 472–484.

Hein, A., Schreieck, M., Wiesche, M., & Krcmar, H. (2016, March). *Multiple-case analysis on governance mechanisms of multi-sided platforms.* Paper presented at Multikonferenz Wirtschaftsinformatik, Ilmenau, Germany.

Henry, A. D. (2011). Ideology, power, and the structure of policy networks. *Policy Studies Journal, 39*(3), 361–383.

Hofstede, G. (1983). National cultures in four dimensions: A research-based theory of cultural differences among nations. *International Studies of Management & Organization, 13*(1–2), 46–74.

Howlett, M., Mukherjee, I., & Koppenjan, J. (2017). Policy learning and policy networks in theory and practice: The role of policy brokers in the Indonesian biodiesel policy network. *Policy and Society, 36*(2), 233–250.

Jeffares, S., & Skelcher, C. (2008). *Democratic subjectivities in network governance: Using web-enabled Q-methodology with European public managers.* Paper presented at the Annual Conference Group for Public Administration, Erasmus University, Rotterdam, Netherlands.

Kamkhaji, J. C., & Radaelli, C. M. (2017). Democratic subjectivities in network governance: Crisis, learning and policy change in the European Union. *Journal of European Public Policy, 24*(5), 714–734.

Katzenbach, C., Hofmann, J., Gollatz, K., Musiani, F., Epstein, D., DeNardis, L., Hackl, A., & Blanchette, J. F. (2015, October). *Doing Internet governance: STS-informed perspectives on ordering the net.* Paper presented at the 16th Annual Meeting of Internet Researchers, Phoenix, Arizona.

Kellogg Foundation. (2004). *Logic model development guide: Using logic models to bring together planning, evaluation, and action.* Retrieved from https://www.bttop.org/sites/default/files/public/W.K.%20Kellogg%20LogicModel.pdf

Klein, H. (2002). ICANN and Internet governance: Leveraging technical coordination to realize global public policy. *The Information Society, 18*(3), 193–207.

Kleinwächter, W. (2000). ICANN between technical mandate and political challenges. *Telecommunications Policy, 24*(6–7), 553–563.

Krahmann, E. (2017). Legitimizing private actors in global governance: From performance to performativity. *Politics and Governance, 5*(1), 54–62.

Lambright, W. H. (1976). *Governing science and technology.* New York, NY: Oxford University Press.

Leach, W. D., Weible, C. M., Vince, S. R., Siddiki, S. N., & Calanni, J. C. (2013). Fostering learning through collaboration: Knowledge acquisition and belief change in marine aquaculture partnerships. *Journal of Public Administration Research and Theory, 24*(3), 591–622.

Levinson, N. S. (2012). Ecologies of representation: Knowledge, networks, & innovation in Internet governance. Paper presented at the American Political Science Association annual meeting. Available at https://ssrn.com/abstract=2108671

Levinson, N. S. (2015). A tri-decennia view of knowledge transfer research: What works in diffusion and development contexts. *Journal of International Communication, 21*(2), 153–168.

Levinson, N. S., & Marzouki, M. (2016). IOs and global Internet governance interorganizational architecture. In F. Musiani, D. Cogburn, L. DeNardis, & N. S. Levinson

(Eds.), *The turn to infrastructure in Internet governance* (pp. 47–72). New York, NY: Palgrave.

Lewis, J. M. (2011). The future of network governance research: Strength in diversity and synthesis. *Public Administration, 89*, 1221–1234.

McFadgen, B., & Huitema, D. (2017). Are all experiments created equal? A framework for analysis of the learning potential of policy experiments in environmental governance. *Journal of Environmental Planning and Management, 60*(10), 1765–1784.

Moyson, S. (2017). Cognition and policy change: The consistency of policy learning in the advocacy coalition framework. *Policy and Society, 36*(2), 320–344.

Moyson, S., Scholten, P., & Weible, C. M. (2017). Policy learning and policy change: Theorizing their relations from different perspectives. *Policy and Society, 36*(2), 161–177.

Musiani, F., Cogburn, D., DeNardis, L., & Levinson, N. S. (Eds.). (2016). *The turn to infrastructure in Internet governance*. New York, NY: Palgrave.

Newig, J., Günther, D., & Pahl-Wostl, C. (2010). Synapses in the network: Learning in governance networks in the context of environmental management. *Ecology and Society, 15*(4), 24.

Newig, J., Kochskämper, E., Challies, E., & Jager, N. W. (2016). Exploring governance learning: How policymakers draw on evidence, experience and intuition in designing participatory flood risk planning. *Environmental Science & Policy, 55*, 353–360.

Nwagwu, W. E., & Iheanatu, O. (2011). Use of scientific information sources by policymakers in the science and technology sector of Nigeria. *African Journal of Library, Archives & Information Science, 21*(1), 59–71.

Nye, J. S. (2014). *The regime complex for managing global cyber activities*. Global Commission on Internet Governance Paper Series (Paper no. 1). Centre for International Governance Innovation/Chatham House. Retrieved from https://www.cigionline.org/sites/default/files/gcig_paper_no1.pdf

Ojo, A., & Mellouli, S. (2018). Deploying governance networks for societal challenges. *Government Information Quarterly, 35*(4), S106–S112.

Or, N. H., & Aranda-Jan, A. C. (2017). The dynamic role of state and nonstate actors: Governance after global financial crisis. *Policy Studies Journal, 45*(S1).

Pahl-Wostl, C. (2009). A conceptual framework for analyzing adaptive capacity and multi-level learning processes in resource governance regimes. *Global Environmental Change, 19*, 354–365.

Provan, K. G., & Kenis, P. N. (2008). Modes of network governance: Structure, management and effectiveness. *Journal of Public Administration Research and Theory, 18*(2), 229–252.

Provan, K. G., & Lemaire, R. H. (2012). Core concepts and key ideas for understanding public sector organizational networks: Using research to inform scholarship and practice. *Public Administration Review, 72*(5), 638–648.

Raab, J., Mannak, R. S., & Cambré, B. (2013). Combining structure, governance, and context: A configurational approach to network effectiveness. *Journal of Public Administration Research and Theory, 25*(2), 479–511.

Raymond, M., & DeNardis, L. (2015). Multistakeholderism: Anatomy of an inchoate global institution. *International Theory, 7*(3), 572–616.

Reagans, R. (2011). Close encounters: Analyzing how social similarity and propinquity contribute to strong network connections. *Organization Science, 22*(4), 835–849.

Rogers, E. (1962). *Diffusion of innovations.* New York, NY: Free Press.

Ruggie, J. G. (2001). Global governance net: The global compact as learning network. *Global Governance, 7*(4), 371–378.

Rutherford, K. R. (2000). The evolving arms control agenda: Implications of the role of NGOS in banning antipersonnel landmines. *World Politics, 53*(1), 74–114.

Schneider, S., & Heinecke, L. (2019). The need to transform science communication from being multicultural via cross-cultural to intercultural. *Advances in Geosciences, 46*, 11–19.

Schreieck, M., Hein, A., Wiesche, M., & Krcmar, H. (2018). The challenge of governing digital platform ecosystems. In C. Linnhoff-Popien, R. Schneider, & M. Zaddach (Eds.), *Digital marketplaces unleashed* (pp. 527–538). Berlin, Germany: Springer.

Spekkink, W. A., & Boons, F. A. (2015). The emergence of collaborations. *Journal of Public Administration Research and Theory, 26*(4), 613–630.

Steijn, B., Klijn, E., & Edelenbos, J. (2011). Public private partnerships: Added value by organizational form or management? *Public Administration, 89*(4), 1235–1252.

Stone, D. (2013). *Knowledge actors and transnational governance: The private-public policy nexus in the global agora.* London, UK: Palgrave Macmillan.

Stone, D. (2019). Transnational policy entrepreneurs and the cultivation of influence: Individuals, organizations and their networks. *Globalizations, 16(2), 1–17.*

Szulanski, G. (2002). *Sticky knowledge: Barriers to knowing in the firm.* New York, NY: Sage.

van Eeten, M. (2017). Patching security governance: An empirical view of emergent governance mechanisms for cybersecurity. *Digital Policy, Regulation and Governance, 19*(6), 429–448.

Visser, M., & van der Togt, K. (2016). Learning in public sector organizations: A theory of action approach. *Public Organization Review, 16*(2), 235–249.

Weber, E. P., & Khademian, A. M. (2008). Wicked problems, knowledge challenges, and collaborative capacity builders in network settings. *Public Administration Review*, *68*(2): 334–349.

Weible, C. M., & Carter, D. P. (2017). Advancing policy process research at its overlap with public management scholarship and nonprofit and voluntary action studies. *Policy Studies Journal*, *45*(1), 22–49.

Weible, C. M., Sabatier, P.A., & McQueen, K. (2009). Themes and variations: Taking stock of the Advocacy Coalition Framework. *Policy Studies Journal*, *37*(1), 121–140.

Weiss, C. (1979). The many meanings of research utilization. *Public Administration Review*, *39*(5), 426–431.

Wurzel, R., Liefferink, D., & Tomey, D. (2019). Pioneers, leaders and followers in multilevel and polycentric climate governance. *Environmental Politics*, *28*(1), 1–21.

Editors

Laura DeNardis is an author and professor and is globally recognized as one of the most read scholars in Internet governance. She is a tenured Professor in the School of Communication at American University in Washington, DC, where she serves as Faculty Director of the Internet Governance Lab. In 2018, she was the recipient of American University's highest faculty award, Scholar-Teacher of the Year. Her six books include *The Internet in Everything: Freedom and Security in a World with No Off Switch* (Yale University Press, 2020), *The Global War for Internet Governance* (Yale University Press, 2014), *Opening Standards: The Global Politics of Interoperability* (MIT Press, 2011), and *Protocol Politics: The Globalization of Internet Governance* (MIT Press, 2009). With a background in information engineering and a doctorate in science and technology studies (STS), she studies the social and political implications of Internet technical architecture and governance. She is an affiliated fellow of the Yale Law School Information Society Project and served as its Executive Director from 2008 to 2011. She is also currently a Senior Fellow at the Columbia University School of International and Public Affairs. Her expertise and scholarship have been featured in *Science Magazine*, the *Economist*, the *New York Times*, *Time*, *Christian Science Monitor*, *Slate*, *Forbes*, the *Atlantic*, and the *Wall Street Journal*, on *National Public Radio*, and by Reuters, among others. She holds an engineering science degree from Dartmouth College, an MEng from Cornell University, a PhD in science and technology studies from Virginia Tech, and she was awarded a postdoctoral fellowship from Yale Law School.

Derrick L. Cogburn is a tenured professor at American University in Washington, DC. He has a joint appointment in the School of International Service, where he serves in the International Communication and International Development Programs, and in the Kogod School of Business, where he serves in the Department of Information Technology and Analytics. He also serves as the founding Executive Director of the AU Institute on Disability and Public Policy and is a Faculty Director of the Internet Governance Lab. He directs the Center for Research on Collaboratories and Technology Enhanced Learning Communities (COTELCO), an award-winning social science research collaboratory investigating the social and technical factors that influence geographically distributed collaborative knowledge work, particularly between developed and developing countries. His research and teaching also includes global information and communication technology and socioeconomic development, multistakeholder institutional mechanisms for Internet governance, and transnational policy networks and epistemic communities. He has published in major journals such as *Telecommunications Policy*, *International Studies Perspectives*, *Journal of International Affairs*, *Assistive Technology*, and *Information Technologies and International Development*. He has published with or advised the Center for Strategic and International Studies, UN World Institute for Development Economics Research, World Bank, UNESCO, International Telecommunication Union, and UN Economic Commission for Africa. He has served as principal investigator or co–principal investigator in externally supported research of over $11 million, with grants from sources as diverse as the National Science Foundation, US Department of Education, JPMorgan Chase, Microsoft Research, the W. K. Kellogg Foundation, and the Nippon Foundation. He is editor of the Palgrave Macmillan book series Information Technology and Global Governance. He is past president of the Information Technology and Politics section of the American Political Science Association and past president of the International Communication section of the International Studies Association. He is a founding member and past vice chair of the Global Internet Governance Academic Network. He holds a bachelor's degree from the University of Oklahoma, and master's and doctoral degrees from Howard University. @derrickcogburn

Nanette S. Levinson is a faculty director of the Internet Governance Lab and a tenured faculty member in the School of International Service (SIS) at American University, where she served as Associate Dean from 1988 to 2005 and from 2015 to 2018. She also serves as Academic Director of the SIS/Sciences-Po Exchange Program. Her research and teaching focus on Internet and global governance, including knowledge transfer and innovation in complex, cross-national, cross-cultural, and cross-organizational systems (such as online settings). She has studied Internet governance since the early days of the Internet Corporation for Assigned Names and Numbers and has been involved in research collaborations with colleagues in France, in Japan, and at American University. She served as the first elected chair of the Global Internet Governance Academic Network. Her leadership positions also include first woman chair of the National Conference on the Advancement of Research, cofounder of the American Society for Public Administration's section on Government and Business, and past president/chair of both the International Studies Association's International Communication section and the American Political Science Association's Information Technology and Politics section. Additionally, she founded and serves as cochair of the Hawaii International Conference on Systems Science's Digital and Social Media Minitrack on Culture, Identity, and Inclusion. Recipient of awards including those for outstanding teaching, program development, honors programming, academic affairs administration, and multicultural affairs, she has designed cocurricular collaborative learning opportunities on campus and research-based training programs for the private and public sectors. In 2011, the Ashoka Foundation presented her with its Award for Outstanding Contributions to Social Entrepreneurship Education. She received her bachelors, masters, and doctorate degrees from Harvard University.

Francesca Musiani (PhD, socioeconomics of innovation, MINES ParisTech, 2012), has been Associate Research Professor at the French National Center for Scientific Research (CNRS) since 2014. She is Deputy Director of the Center for Internet and Society of CNRS, which she cofounded in 2019. She is also an associate researcher at the Center for the Sociology of Innovation (i3/MINES ParisTech) and a Global Fellow at the Internet Governance Lab of American University in Washington, DC. Since 2006 her research work has focused on Internet governance, in an interdisciplinary perspective that merges information and communication sciences, science and technology studies (STS), and international law. Her most recent research explores the development and use of encryption technologies in secure messaging (European Commission's Horizon 2020 project NEXTLEAP, 2016–2018), digital resistances to censorship and surveillance in the Russian Internet (French National Research Agency, or ANR, project ResisTIC, 2018–2021), and the governance of web archives (ANR project Web90, 2014–2017 and CNRS Attentats-Recherche project ASAP, 2016). Her theoretical work explores STS approaches to Internet governance, with particular attention paid to socio-technical controversies and to governance by architecture and by infrastructure. Her most recent book is *Qu'est-ce qu'une archive du Web?* (What is a web archive?, OpenEdition Press, 2019), with C. Paloque-Bergès, V. Schafer, and B. Thierry, recipient of the OpenEdition Books Select distinction. She is academic editor for *Internet Policy Review*. She is vice president for research of the Internet Society France. Since 2017, she has cochaired the Communication Policy and Technology section of the International Association for Media and Communication Research, after having led its emerging scholars network (2012–2016). @franmusiani / https://cis.cnrs.fr/francesca-musiani

Contributors

Farzaneh Badiei is a research scholar at Yale Law School, leading the Social Media Governance Initiative. Prior to that she was a research associate at the Georgia Institute of Technology, School of Public Policy, and the Executive Director of the Internet Governance Project. For nearly a decade, she has been a part of the Internet governance research and professional community. She has conducted research at the Humboldt Institute for Internet and Society and the Syracuse School of Information Studies. She received her PhD from the University of Hamburg, Institute of Law and Economics.

Davide Beraldo is a postdoctoral researcher in the DATACTIVE and ALEX projects, and a Lecturer in New Media and Digital Culture at the Department of Media Studies, University of Amsterdam. He holds a PhD (cum laude) in sociology from the University of Amsterdam and the University of Milan, and a master (cum laude) in social sciences. He is currently working on investigating political biases in recommendation systems of popular social media and on developing a Social Movement Studies framework for the conceptualization of data activism. In his PhD dissertation, he explored the epistemological and methodological implications of the digital mediation of social movements, investigating large datasets of social media data related to the Occupy and Anonymous protest movements. He has a background in computer programming and has worked as a freelance developer and consultant for political and market research agencies. His research interests include digital sociology, social movements, algorithms, online networks, and epistemology of complexity.

Sandra Braman is Professor of Communication and Abbott Professor of Liberal Arts at Texas A&M University. Her books include *Change of State: Information, Policy, and Power* (currently undergoing revision for a second edition) and the edited volumes *Biotechnology and Communication: The Meta-technologies of Information*; *The Emerging Global Information Policy Regime*; and *Communication Researchers and Policy-Making*, and she has written almost 100 scholarly journal articles and book chapters. She is editor of the Information Policy Series at MIT Press and Fellow of the International Communication Association. She is former chair of the Communication Law and Policy Division of the International Communication Association and former head of the Law Section of the International Association of Media and Communication

Research. Her research has been funded by the Rockefeller Foundation, Ford Foundation, Soros Foundation, and the US National Science Foundation.

Ronald J. Deibert is Professor of Political Science and Director of the Citizen Lab at the Munk School of Global Affairs and Public Policy, University of Toronto. The Citizen Lab undertakes interdisciplinary research at the intersection of global security, information and communication technologies, and human rights. The research outputs of the Citizen Lab are routinely covered in global media, including over two dozen reports receiving front page coverage in the *New York Times*, *Washington Post*, and other media over the last decade. He is the author of *Black Code: Surveillance, Privacy, and the Dark Side of the Internet* (Random House, 2013) and numerous books, chapters, articles, and reports on Internet censorship, surveillance, and cybersecurity. In 2013 he was appointed to the Order of Ontario and awarded the Queen Elizabeth II Diamond Jubilee medal, for being "among the first to recognize and take measures to mitigate growing threats to communications rights, openness and security worldwide."

Wendy Hall, DBE, FRS, FREng, is Regius Professor of Computer Science at the University of Southampton, UK, and an executive director of the Web Science Institute at Southampton. Her influence as one of the first to undertake serious research in multimedia and hypermedia has been significant in many areas, including digital libraries, the development of the Semantic Web, and the emerging discipline of web science. She became a Dame Commander of the British Empire in 2009 and is a fellow of the Royal Society. She has been president of the Association for Computing Machinery, senior vice president of the Royal Academy of Engineering, and a member of the UK Prime Minister's Council for Science and Technology. She was a founding member of the European Research Council and chair of the European Commission's Information Society Technologies Advisory Group, a member of the Global Commission on Internet Governance, and a member of the World Economic Forum's Global Futures Council on the Digital Economy. Dame Wendy was cochair of the UK government's review of artificial intelligence, *Growing the Artificial Intelligence Industry in the UK* (2017), and became the UK government's first Skills Champion for AI in the UK in 2018.

Jeanette Hofmann is a political scientist with a focus on Internet regulation. At the WZB Berlin Social Science Center she heads the research group Politics of Digitalization, which studies how today's societies make sense of and shape the digital transformation. The group examines processes of digitalization both as a resource of political governance (regulation through digitalization) and as an object of political decision-making (regulation of digitalization). She is Professor of Internet Politics at the Freie Universität Berlin, Director and Founder of the Berlin-based Alexander von Humboldt Institute for Internet and Society, and principal investigator at the newly founded Weizenbaum Institute for the Networked Society. At the latter institute, she heads two research groups, one on digitalization and democracy and one on quantification and regulation. Her two current research foci are platform governance and democratic change. In addition to her academic work, she has been involved in various political processes such as the UN World Summit on the Information Society and the Internet Governance Forum on the international level and, as an expert member, in the Internet and Digital Society committee of inquiry of the German Parliament on the national level. At present she is a member of the expert group to the EU Observatory on the Online Platform Economy. She also heads two national academic expert commissions, one on digitalization and democracy (Academy of Sciences Leopoldina) and one on youth engagement in the digital age (German government).

Eric Jardine is an assistant professor of political science at Virginia Tech and a fellow at the Centre for International Governance Innovation. His research focuses on the uses and abuses of the Dark Web, the measurement of trends in cybercrime data, and the politics surrounding anonymity-granting technologies and encryption. His work has been published in a number of peer-reviewed outlets, including *New Media & Society*, *International Journal of Drug Policy*, *Journal of Cyber Policy*, *First Monday*, *Intelligence and National Security*, *Terrorism and Political Violence*, and *Studies in Conflict and Terrorism*. He is the coauthor, with Fen Hampson, of *Look Who's Watching: Surveillance, Treachery and Trust Online* (CIGI/MQUP, 2017).

Rikke Frank Jørgensen is a Senior Researcher at the Danish Institute for Human Rights in Copenhagen. Her research focuses on the intersection between human rights and technology and covers issues such as the role of private actors in the online domain, Internet users' human rights, and Internet regulation and governance. Besides her scholarly activities, she has served as an adviser to the Danish government, participated in the Council of Europe's Committee on Human Rights for Internet Users, and been closely involved in civil society networks such as European Digital Rights. Her most recent book (as editor), *Human Rights in the Age of Platforms* (MIT Press, 2019) examines the human rights implications of the social web, through the lens of datafication, platforms, and human rights regulation.

Aastha Madaan is a senior data scientist in the Advanced Digital Engineering group at Arup, London. She is responsible for leading the technical delivery of data-driven products and analytical and machine learning projects for the aviation, infrastructure, and cities sectors. She is a visiting research fellow at the School of Electronics and Computer Science, University of Southampton, working on design of a data sharing infrastructure for cognitive IoT ecosystems. Previously she worked as a research fellow at the University of Southampton on research problems and design of an IoT test bed for secure data sharing, innovation, privacy, and data security. She received her PhD in computer science from the Database Laboratory at the University of Aizu, Fukushima, Japan, in 2014 and has a master's degree in computer science from the University of Delhi, India (2006–2008). Her research interests include data innovation in cognitive IoT, data science, and emerging technologies such as edge computing and artificial intelligence.

Stefania Milan (stefaniamilan.net) is Associate Professor of New Media and Digital Culture at the University of Amsterdam. Her research explores the intersection of digital technology, governance, and activism, with emphasis on critical data practices and autonomous infrastructure. She enjoys creating bridges between research, activism, and policy making, and is passionate about methodological innovation. Her work has received funding from, among others, the European Research Council and the Dutch Research Council (NWO). Stefania holds a PhD in political and social sciences from the European University Institute. Prior to joining the University of Amsterdam, she worked at the Citizen Lab (University of Toronto), Tilburg University, Central European University, the University of Lucerne (Switzerland), and the Robert Schuman Center for Advanced Studies at the European University Institute. Stefania is the author of *Social Movements and Their Technologies: Wiring Social Change* (Palgrave Macmillan, 2013) and coauthor of *Media/Society* (Sage, 2011). In 2019, she coedited a special issue of the journal *Policy and Internet* dedicated to Internet Infrastructure and Human Rights. As a digital rights advocate, she has been vocal in Internet governance and cybersecurity circles.

Milton L. Mueller is Professor in the School of Public Policy at the Georgia Institute of Technology. He is an internationally prominent scholar specializing in the political economy of information and communication. The author of seven books and scores of journal articles, his work informs public policy, science and technology studies, law, economics, communications, and international studies. His books *Will the Internet Fragment? Sovereignty, Globalization and Cyberspace* (Polity, 2017), *Networks and States: The Global Politics of Internet Governance* (MIT Press, 2010), and *Ruling the Root: Internet Governance and the Taming of Cyberspace* (MIT Press, 2002) are acclaimed scholarly accounts of the global governance regime emerging around the Internet. He is the cofounder and director of the Internet Governance Project, which has played a prominent role in shaping global Internet policies and institutions such as ICANN and the Internet Governance Forum. He has participated in proceedings and policy development activities of ICANN, the International Telecommunication Union, and the US National Telecommunications and Information Administration (NTIA) and regulatory proceedings of the European Commission, China, Hong Kong, and New Zealand. He has served as an expert witness in prominent legal cases related to domain names and telecommunication policy. He was elected to the Advisory Committee of the American Registry for Internet Numbers from 2013 to 2016 and appointed in 2014 to the Internet Assigned Numbers Authority Stewardship Coordination Group. He helped create the Global Internet Governance Academic Network, an international association of scholars focused on Internet governance.

Kieron O'Hara is an associate professor in electronics and computer science at the University of Southampton, UK. His interests are in the philosophy and politics of digital modernity, particularly the World Wide Web; key themes are trust, privacy, and ethics. He is the author of several books on technology and politics; the latest, with Nigel Shadbolt, David De Roure, and Wendy Hall, *The Theory and Practice* *of Social Machines* (Springer), appeared in 2019. He has also written extensively on political philosophy and British politics. He is one of the leads on the UK Anonymisation Network (UKAN), which disseminates best practice in data anonymization.

Niels ten Oever is a PhD candidate with the DATACTIVE Research Group at the Department of Media Studies, University of Amsterdam, and affiliated with its Political Science Department. His research focuses on how values, like human rights, are inscribed in the Internet infrastructure through its transnational governance. He seeks to understand how invisible infrastructures provide a socio-technical ordering to our societies and how that might influence the distribution of wealth, power, and possibilities. Prior to starting his PhD, he worked as Head of Digital for the international freedom of expression not-for-profit ARTICLE19, where he designed, fundraised, and set up the digital program that covered the Internet Engineering Task Force, the Internet Corporation for Assigned Names and Numbers, the Institute for Electrical and Electronic Engineers, and the International Telecommunication Union. He also designed and implemented freedom of expression projects with Free Press Unlimited. He holds a cum laude MA in philosophy from the University of Amsterdam.

Rolf H. Weber is Professor of International Business Law at Zurich University; there he acts as codirector of the Center for Information Technology, Society, and Law and as codirector of the Blockchain Center. Furthermore, he is practicing attorney in one of the largest independent Swiss law firms in Zurich. From 2000 to 2015 he was Visiting Professor at the University of Hong Kong. He is a member of the editorial boards of several Swiss and international legal periodicals, a member of SIEL/AIELN from their beginnings, and a member of the European Dialogue on Internet Governance. He is fluent in German, English, and French. His main fields of research and practice are IT and the Internet, international business competition, and international trade and finance law. He publishes and speaks regularly on Internet-related legal issues.

Index